Rigor, Relevance, and Relationships

Making Mathematics Come Alive with Project-Based Learning

Edited by
Jean Lee
University of Indianapolis, Indiana
and
Enrique Galindo
Indiana University Bloomington

NATIONAL COUNCIL OF
TEACHERS OF MATHEMATICS

www.nctm.org/more4u
Access code: RRR14798

Copyright © 2018 by
The National Council of Teachers of Mathematics, Inc.
1906 Association Drive, Reston, VA 20191-1502
(703) 620-9840; (800) 235-7566; www.nctm.org
All rights reserved

Library of Congress Cataloging-in-Publication Data

Names: Lee, Jean (Jean Sangmin), author. | Galindo, Enrique (Enrique, Galindo-Morales), author.
Title: Rigor, relevance, and relationships : Making mathematics come alive with project-based
learning / Jean Lee and Enrique Galindo. Description: Reston, VA : National Council of Teachers
of Mathematics, [2017] | Includes bibliographical references. Identifiers: LCCN 2017052243 |
ISBN 9780873537704 Subjects: LCSH: Mathematics--Study and teaching--Standards--United
States. | Project method in teaching--United States.   Classification: LCC QA13 .L4394 2017 |
DDC 510.71/073--dc23 LC record available at https://lccn.loc.gov/2017052243

The National Council of Teachers of Mathematics supports and advocates for the
highest-quality mathematics teaching and learning for each and every student.

When forms, problems, or sample documents are included or are made available on NCTM's website, their use
is authorized for educational purposes by educators and noncommercial or nonprofit entities that have purchased
this book. Except for that use, permission to photocopy or use material electronically from *Rigor, Relevance, and
Relationships: Making Mathematics Come Alive with Project-Based Learning* must be obtained from www.copyright.
com or by contacting Copyright Clearance Center, Inc. (CCC), 222 Rosewood Drive, Danvers, MA 01923,
978-750-8400. CCC is a not-for-profit organization that provides licenses and registration for a variety of users.
Permission does not automatically extend to any items identified as reprinted by permission of other publishers or
copyright holders. Such items must be excluded unless separate permissions are obtained. It is the responsibility of
the user to identify such materials and obtain the permissions.

The publications of the National Council of Teachers of Mathematics present a variety of viewpoints. The views
expressed or implied in this publication, unless otherwise noted, should not be interpreted as official positions of
the Council.

The following forms are adapted from Buck Institute for Education, www.bie.org: Project Planning Form, Scaffolding
NTKs: Activity and Assessment Planning, Project Calendar, and Six Problem-Solving Phases Planning Worksheet.

Printed in the United States of America

Table of Contents

Preface

Rigor, Relevance, and Relationships: Making Mathematics Come Alive with Project-Based Learning is divided into four sections. In section I, we introduce a teaching methodology called *project-based learning* (PBL). We describe the nuts and bolts of designing an effective PBL unit (chapter 1), outline the importance of college and career readiness and how PBL instruction can play a key role in preparing students for these future challenges (chapter 2), and present a literature review on the effectiveness of PBL (chapter 3). Sections II and III showcase PBL units that have been designed and implemented by high school mathematics teachers in their classrooms. The final section includes tips and resources from mathematics teachers and educators who have taught and researched in PBL settings.

Each unit includes tips on how to sustain **rigor**, engage learners in **relevant** learning, and foster students' **relationships** with their peers and with members of the community. The essential elements of PBL and the connections to the Common Core State Standards for Mathematics (CCSSM) are discussed in detail. Units include a variety of supplemental materials, for example:

- A Project Planning Form and Project Calendar containing details about the project
- Rubrics used throughout the project
- Written and digital examples of critical points in the PBL process, such as the launch of the project, establishment of effective small-group norms, and project presentations
- Examples of student work and student learning
- Other resources that help make PBL effective in math classrooms

Supplementary digital resources for both detailed and brief PBL units can be found on the National Council of Teachers of Mathematics (NCTM) website (www.nctm.org/more4u).

The detailed PBL units presented in section II are as follows:

- In *Solar Cooking with Conics* (chapter 4), students are challenged to build a solar cooker that uses the sun's rays as the only means of heating a pot of soup. Learners are engaged in becoming more globally aware as they use a number of digital tools to meet this mathematical challenge.
- *Interest in Interest* (chapter 5) uses the idea of shopping for a new car as a framework for students' learning about exponential growth, decay, and logarithms. Students learn about financial literacy—a real-life skill they will need in the near future.
- The *Polyhedron Tiling Artwork Project* (chapter 6) incorporates art into students' work in geometry. It is a great example of how to scaffold non-PBL learners' experiences by implementing some PBL practices in a mathematics classroom.
- *I'm Hair to Help* (chapter 7) is a service learning project in which students are introduced to the Locks of Love organization; in addition to math skills, students learn about and practice empathy.

The brief PBL units presented in section III are as follows:

- *To Netbook or Not to Netbook?* (chapter 8) is based on an actual situation: investigating whether the one-to-one Netbooks provided by a school were being used effectively. The unit empowered learners to gather data to help the school administration.
- *"Worth"-Quake Insurance* (chapter 9) focuses on a recent series of earthquakes that alarmed a community. Learners use exponential growth graphs to make a compelling argument for purchasing earthquake insurance, in the process developing a relationship with a popular local insurance agency.
- *Super Baugh I: Flacco vs. Kaepernick* (chapter 10) draws on learners' interest in the Super Bowl and engages them in exploring and modeling phenomena to predict the outcome of athletes' performances.
- In *Don't Sweat It!* (chapter 11), students learn about data analysis and mathematical modeling through research on healthy eating habits and exercise routines.

Readers are encouraged to adapt these units and use them in their own instructional settings. We also encourage readers to design and implement new PBL units for their own use and to share them with the broader community of mathematics educators. The templates in chapter 1 (Project Planning Form, Scaffolding NTKs [Need-to-Knows], Project Calendar, Six Problem-Solving Phases Planning Worksheet) are provided to aid teachers in designing or adapting PBL units for their own classrooms.

About More4U Materials

Many of the teachers who contributed PBL units to this book were generous enough to share some of their materials, including lesson plans, slideshows, student work samples, quizzes, reflection prompts, handouts, and assessment rubrics. These materials are included to give readers a better sense of the scaffolds and resources these teachers used when implementing their PBL units. Teachers interested in implementing the units may use or adapt these materials for their own classrooms.

more**4U**

More4U materials can be found on the NCTM website (www.nctm.org/more4u). Look for the More4U icon throughout this book for additional resources.

List of More4U Resources by Chapter

Chapter 1

1. Project Calendar Template (Microsoft Word)
2. Project Planning Form Template (Microsoft Word)
3. Scaffolding NTKs Template (Microsoft Word)
4. Six Problem-Solving Phases Planning Worksheet Template (Microsoft Word)

Chapter 4

1. Challenge 3 Guiding Questions (Microsoft Word)
2. Challenge 3 (GeoGebra)
3. Challenge Packet (Microsoft Word)
4. Conic Sections Worksheet (Microsoft Word)
5. Conics (GeoGebra)
6. Family Letter (Microsoft Word)
7. Final Quiz (Microsoft Word)
8. Focal Length of Solar Cooker (GeoGebra)
9. Sample Student Contract (Microsoft Word)
10. Sample Student Work (PDF)
11. Sample Warm-Up (Microsoft Word)
12. Take-Home Quiz (Microsoft Word)
13. Teacher Testimony (movie)

Chapter 5

1. Affordability Workshop Lesson Plan (Microsoft Word)
2. Car Buying Workshop Lesson Plan (Microsoft Word)
3. Car Depreciation Worksheet (Microsoft Word)
4. Group Average Car Budget Worksheet (Microsoft Word)
5. Group Contract (Microsoft Word)
6. Individual Car Budget Worksheet (Microsoft Word)

Chapter 6

1. Apothems Worksheet (Microsoft Word)
2. Math Lab: Painting a House (Microsoft Word)
3. Math Lab: Surface Area Applications (Microsoft Word)
4. Math Lab: Surface Area vs. Volume (Microsoft Word)
5. Polyhedra and Euler's Theorem Worksheet (Microsoft Word)
6. Quiz (Microsoft Word)
7. Sample Student Slideshow (Microsoft Word)
8. Similar Solids Worksheet (Microsoft Word)
9. Surface Area and Volume of Spheres Worksheet (Microsoft Word)
10. Surface Area of Prisms (etc.) Worksheet (Microsoft Word)
11. Test (Microsoft Word)
12. Volume of Prisms (etc.) Worksheet (Microsoft Word)
13. Word Bank (Microsoft Word)

Chapter 8

1. Day 1 Detailed Lesson Plan (Microsoft Word)
2. Final Quiz (Microsoft Word)
3. Socratic Discussion (PowerPoint)

Chapter 10

1. Guidance for Group Presentations (Microsoft Word)

Acknowledgments

We express sincere thanks and gratitude to Catherine A. Brown, who helped with the conceptualization of this book. Her insights and perspectives helped shape our efforts to make this book as useful as possible for secondary math PBL teachers.

We also extend our appreciation to teachers and facilitators who have used PBL methods and found them successful in their classrooms. Your tireless endeavors on this PBL journey are not forgotten; and your commitment to bringing rigor, relevance, and relationships into a math classroom is inspiring.

An Inquiry-Based Approach: Project-Based Learning

Jean Lee, *University of Indianapolis, Indiana*

The project-based learning (PBL) model is based on the assumption that most academic content is learned best in the context of projects. A PBL curriculum engages learners in meaningful problems that are important to them while advancing their creativity and problem-solving abilities.

PBL is an inquiry-based instructional approach that reflects a learner-centered environment and concentrates on learners' application of disciplinary concepts, tools, experiences, and technologies to research the answers to questions and solve real-world problems (Krajcik and Blumenfeld 2006; Markham, Larmer, and Ravitz 2003). PBL can help increase both the range of learners' interests and their conceptual understanding of mathematics content. Teachers support ways for learners to construct their own understanding and orchestrate conversations in which learners explore complex connections and relationships among ideas.

General core principles and practices of PBL include the following:

- A professional culture of trust, respect, and responsibility among the learners and the teacher
- A focus on 21st century skills and academic standards, such as the Common Core State Standards for Mathematics (CCSSM)
- Scaffolding activities that include student-centered instruction to increase relevance and rigor
- Learning connected to other subject areas and to the post–high school world, for college and career readiness
- Infusion of technology as a tool for communicating, collaborating, and learning
- Partnerships with community institutions, such as higher education, businesses, and nonprofit agencies, so that learners can build relationships with other local stakeholders

In this book, we showcase a number of PBL units that were designed and implemented by mathematics teachers, coupled with tips and narratives to support readers in implementing PBL.

"Doing Projects" Compared to PBL

In a traditional math classroom, "doing projects" happens at the end of a unit, after the teacher has presented the content through a series of lessons and learners have completed homework assignments, practice problems, readings, lectures, textbook activities, and class discussions. Learners then demonstrate their understanding of the content in a culminating project (see fig. 1.1).

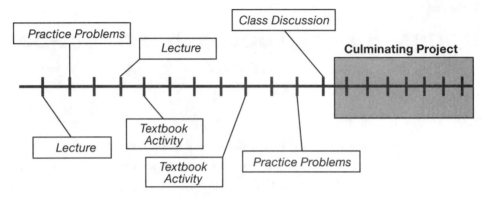

Fig. 1.1. Timeline of "doing projects."

Because PBL instruction occurs through an "extended inquiry process" (Markham, Larmer, and Ravitz 2003, p. 4), the project does not take place at the end of the unit as a culminating product. Instead, it is given to learners at the beginning of a unit specifically to engage them in the content.

The project is launched with an *Entry Event*.

[handwritten margin note: What does "extended" look like in MS? How to build this?]

Entry Event: The activity used to "kick off" the PBL unit (e.g., a letter, video, or other presentation of a real-life problem) to maximize learners' engagement and inquiry.

A *Driving Question*, presenting an authentic problem, then pulls or guides learners through the curriculum, giving them a further incentive to learn the mathematics content (see fig. 1.2).

Driving Question: An open-ended challenge or problem that learners explore throughout the project.

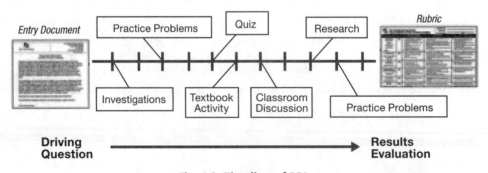

Fig. 1.2. Timeline of PBL.

Instruction is integrated into the project as learners need it, thus increasing the relevance of the mathematics and the need to learn it.

Many elements found in a traditional classroom—such as practice problems, textbook activities, and class discussions—are integrated into the unit in response to learners' *Need-to-Knows (NTKs)*, based on the Entry Event.

Need-to-Knows (NTKs): A list of questions that learners generate after the Entry Event, which pertain to the knowledge and skills needed to successfully complete the project. The list is revisited daily, and more NTKs are added as needed.

However, instead of listening to lectures, learners investigate, build, and research concepts as part of inquiry-based instruction.

Key Features of PBL

Krajcik and Blumenfeld (2006) describe five key features of the PBL model:

- The project or unit starts with a Driving Question that addresses an authentic concern, in order to sustain learners' interest during the project.

- Learners investigate the Driving Question through problem solving, and in the process they learn and apply important ideas specific to the discipline.

- There is collaboration among learners, teachers, and community members to find solutions to the Driving Question.

- Projects often require instructional tools such as technology to enhance learners' learning and their ability to complete the project.

- Learners create tangible products that address the Driving Question and present them to their class and community members.

Designing Effective PBL Units

A number of elements are essential to the design of an effective PBL unit, including a focus on 21st-century skills, an Entry Event, a Driving Question, NTKs, the Project Planning Form, the Project Calendar, scaffolding instruction, and a project rubric. Each is described in greater detail below.

21st-Century Skills

In today's rapidly changing economy and an increasingly tech-savvy industry, employers seek graduates who can solve problems, think critically, exercise creativity, and work in teams (KnowledgeWorks Foundation 2009; Partnership for 21st-century Learning n.d.). These "soft" skills, which are imperative for jobs in the 21st century, are often not addressed in traditional methods of teaching (Markham, Larmer, and Ravitz 2003).

[handwritten margin note: if they are, then what's the benefit of PBL?]

In a PBL environment, learners need to do much more than simply remember information: They need to use higher-order thinking skills, and they must learn to work as a team and contribute to a group effort. They must listen to others, make their own ideas clear when speaking, be able to read a variety of materials, write or otherwise express themselves in various modes, and give effective presentations. These skills, competencies, and habits of mind are known as *21st-century skills*, which include communication, creativity, use of technology, group process and collaboration, problem solving and critical thinking, and task- and self-management (Markham, Larmer, and Ravitz 2003, pp. 25–27). Teaching and assessing these skills helps prepare learners to thrive in today's global economy.

Learners also work independently and take responsibility when they are asked to make choices. Opportunities to make choices and to express their learning in their own voice also help increase learners' educational engagement.

The Project Planning Forms showcased in this book allow readers to see which 21st-century skills are encouraged from the project work or explicitly taught and assessed. (This form is discussed in more detail below.)

Entry Event

PBL units are most often launched with an Entry Event that contextualizes the problem and motivates learners to engage in the content. Entry Events may be letters, documents, videos, presentations, or any other activity that engages learners by presenting an authentic problem.

There are several ways to launch an Entry Event, for example:

- A community member poses a problem to the learners and asks for their help
- Learners watch a video that provides a context for the problem
- Learners read a letter from someone who outlines the expectations for the project and will later evaluate their products

Ideally, the problem presented is an authentic one, and a representative of the company or organization in question can present the challenge, either virtually or in person.

Some teachers have found it expedient to draft the Entry Event letter themselves and then have the community partner approve the draft before signing it. In most cases, partners were happy with this arrangement and did not request any changes.

An Entry Event should accomplish four things:

- Hook the learners
- Allow learners to discern their roles
- Lay out the project or problem to be completed or solved
- Provide information that motivates the learners to ask questions and seek answers

Each PBL unit in this book showcases how teachers used an Entry Event to launch their project and describes how the Entry Event was presented to the learners.

Driving Question

The Driving Question is an open-ended challenge or problem that focuses learners' work and deepens their learning by centering on a significant real-life question, problem, issue, and/or debate. It requires the teacher to articulate a scenario that can be meaningful to learners.

Once learners are engaged by the Entry Event, they should be able to articulate the Driving Question, with the teacher's guidance. (It is also okay to present the Driving Question to the students, particularly for teachers doing PBL for the first time.) Teachers can guide learners to define the problem stated within the Driving Question by having them reflect on the following framework:

> How do we, as . . . (learner's role),
> create/research/develop . . . (task)
> so that . . . (desired outcome)?

For example, the Driving Question presented in *Interest in Interest* (chapter 5) is:

> *How can we, as recent college graduates, determine*
> *the best vehicle to purchase based on our incomes?*

The learners conducting research for this PBL unit are in an Algebra 2 class in an early college career setting. By framing their role as recent college graduates seeking to purchase a vehicle and having them work with actual bank and loan officers, there is a dual layer of authenticity and adult connections made for the learners.

Need-to-Knows (NTKs)

The learners' next task is to generate and record a list of items they will "need to know" or understand in order to answer the Driving Question and ultimately complete the project successfully. Creating this list is an essential problem-solving skill that allows learners to have a voice in how they will receive the content and what kinds of instruction they will need, thus empowering them to take charge of their learning and creating relevance for their learning throughout the project.

A teacher can use a variety of resources to organize the NTKs, including sticky notes, Google Docs, and a shared digital wall (such as Padlet or Wallwisher). The NTK list becomes a living document for the duration of the project: Learners revisit the list daily and add to or revise it to assess their progress.

Before launching the unit, the teacher should anticipate some likely NTKs that learners will have. Each project showcased in this book contains a list of NTKs that teachers anticipated their learners would raise, how they planned to support learners in addressing or answering them, and how this learning would be assessed.

The NTKs listed in each PBL unit are not exhaustive; they are merely examples to indicate the level of detail required in the PBL unit-planning process.

Template 1.1: Scaffolding NTKs: Activity and Assessment Planning (see fig. 1.3) is a tool for teachers to list likely NTKs, possible assignments or activities for learners to demonstrate the NTKs, possible assessments to confirm that learners understand the NTKs, and what learning outcomes will be addressed in the unit (Larmer, Ross, and Mergendollar 2009).

Scaffolding NTKs: Activity and Assessment Planning			
Anticipated Knowledge and Skills Students Need (NTKs)	Assignment or Activity to Address NTKs	How Assignment or Activity Will Be Assessed	Learning Outcomes Addressed in Assignment/Activity

Fig. 1.3. A snapshot of Template 1.1.

While lessons and NTKs are generally planned by the project teacher, they may also be provided by other teachers, experts, mentors, and/or community members, depending on the context of the project. For example, if students are engaging in a service learning project, the community partner will likely need to cover some important content. Partners may also contribute to how students' work will be assessed—what students need to do to successfully complete the culminating products and performances.

Templates can be found at the end of this chapter.

Electronic copies of each template are available on the NCTM website (www.nctm.org/more4u).

Project Planning Form

There are many logistics to consider when planning a PBL unit. Template 1.2: Project Planning Form (adapted from Larmer, Ross, and Mergendollar 2009, pp. 120–121) can guide teachers in laying out their thinking—about the big picture as well as some important details that go into a PBL unit, including its rigor, relevance, and relationships. This form can also help ensure that the Entry Event elicits the NTKs that teachers anticipate from learners. When the NTKs are in alignment with the targeted standards and skills, learners are encouraged to take a deeper dive into their project research.

Each PBL unit showcased in this book includes a completed Project Planning Form.

Project Calendar

The Project Calendar helps teachers plan the scope and sequence of what mathematics is taught and what learning opportunities learners will engage themselves in throughout the unit. However, because PBL learning is a dynamic process, teachers must be flexible about revising the calendar as needed—sometimes daily! PBL teachers often allot one or two "buffer days" in the calendar that are used to answer unanticipated NTKs and to provide extra work time as needed.

Each PBL unit in this book includes a completed Project Calendar. (See Template 1.3 Project Calendar)

Scaffolding Instruction

The ways in which the teacher supports learners' learning (e.g., practice problems, textbook activities, class discussions, investigations, research) and the activities that support the problem-solving process are referred to as *scaffolding techniques*. Scaffolds are integrated into the instruction of the unit as learners need the information, so that their learning is both authentic and relevant. PBL teachers often review the Entry Event, Project Planning Form, and Project Rubric (see below) to ensure that the standards and skills addressed are consistent and that there are no surprises in the expectations across documents for learners.

It is important to balance learning about the *context* of the project with learning about its *content*. For example, in the *Solar Cooking with Conics* unit (chapter 4), learners need to construct a functional solar cooker (*context*) and also master the concepts of conic sections (*content*).

Because PBL involves an "extended process of inquiry" (Markham, Larmer, and Ravitz 2003, p. 4), it is important that learners have opportunities to construct something new—an idea, an interpretation, or a new way of displaying what they have learned. In a PBL mathematics classroom, learners' actual creation of the project's products and the *process* of creating the products are equally important. Mathematics learners should practice problem-solving skills in which they "make sense of problems and persevere in solving them," which is a Mathematical Practice Standard (National Governors Association and Council of Chief State School Officers 2010). Furthermore, NCTM (2009) advocates instructional approaches that make reasoning and sense-making foundational to the mathematics content, as such approaches best prepare learners for citizenship, the workplace, and further study.

To provide learners with these opportunities, teachers must intentionally design math PBL units that mirror the six phases of the problem-solving process:

- Phase 1: What is the need or problem?
- Phase 2: What needs to be in our solution?
- Phase 3: What are possible solutions?
- Phase 4: Which solution should we use?
- Phase 5: How do we create, run, and inspect our solution?
- Phase 6: Reflect back: Did we solve the problem? Do we have a new problem?

Consideration of each phase supports the teacher in explicitly stating how the unit allows learners to practice problem-solving and reasoning skills.

Each PBL unit showcased in this book includes a completed worksheet. (See Template 1.4: Six Problem-Solving Phases Planning Worksheet.)

Project Rubric

Rubrics help learners understand the expectations of the project and prepare them for how they will demonstrate their learning for public scrutiny and critique. PBL unit rubrics are designed so that learners demonstrate not only content mastery but also "soft" skills (i.e., 21st-century skills).

Even though a number of "soft" skills may be encouraged throughout a unit, Larmer, Ross, and Mergendollar (2009) recommend that novice PBL practitioners identify no more than two that are explicitly taught and assessed as project outcomes.

Each PBL project in this book includes the rubric created and used by the teacher.

Roots of PBL

Dewey, Piaget, Vygotsky, and other progressive educators laid the curricular and psychological foundations for PBL instruction. Dewey (1902) observed that children must be guided and provided with appropriate learning experiences if they are to develop a habit of critical examination and inquiry. Piaget (1970) and Vygotsky (1978) further strengthen this method of instruction by focusing on student-centered learning and construction of knowledge through practice and reflection. Student-centered learning, hands-on learning, and guided learning are some of the core values of PBL instruction.

Building on the work of these educators, Krajcik and Blumenfeld (2006) propose four major learning-science ideas that describe the curricular and psychological foundations for PBL instruction: (1) active construction, (2) situated learning, (3) social interactions, and (4) cognitive tools.

Active Construction

According to Krajcik and Blumenfeld (2006), learners must actively construct meaning based on their experiences and interactions in the world in order for deep understanding to occur. Development of understanding is an iterative process in which learners reconstruct what they know from prior experiences and apply it to new experiences and ideas. Thus, rather than passively take in information from the teacher, learners "actively build knowledge as they explore the surrounding world, observe and interact with phenomena, take in new ideas, make connections between new and old ideas, and discuss and interact with others" (p. 319). In a PBL setting, learners construct their knowledge by engaging in real-world activities similar to the kinds of activities that are demanded of experts in the field, such as solving problems and developing artifacts.

While learners actively construct their solutions in PBL settings, the teacher scaffolds content and activities in order to enhance learners' skills and capabilities. Hence, the teacher serves as a facilitator or "metacognitive coach" and designs units so that the learning environment supports and challenges the learners' thinking (Gallagher and Stepien 1996, p. 261). The teacher plays a crucial role in monitoring and assessing each group's and/or learner's progress during PBL activities. During each stage of the problem-solving process, the teacher may interject with additional resources to help learners in their pursuit of a solution to the problem.

Situated Learning

In situated learning, the context of learning is inextricably tied to the situation or context in which the learners are placed (Barab and Duffy 2000; Brown, Collins, and Duguid 1989; Resnick 1987). Rather than present information that learners may or may not be able to use to solve the stated problem, situated learning stresses that knowledge should be presented in context, preferably in a problem-solving scenario. In addition, if teachers want learners to solve complex real-world problems, they need to provide learning opportunities in those contexts (Goldman, Hasselbring, and the Cognition and Technology Group at Vanderbilt 1996; Plucker and Nowak 2000).

Is this what I really think is best for my students?

Situated learning allows learners to acquire information in a meaningful context and relate it to their prior knowledge and experiences so that they can form connections between new information and their prior knowledge. In a PBL setting, teachers design units that anchor all learning activities to a larger task or problem. Contextualizing the learning in this way enables learners to easily see the value and purpose of the tasks and activities they are asked to do (Krajcik and Blumenfeld 2006).

Social Interaction

Vygotsky (1962) notes that learning is a social activity that takes place within the context of a unit. In PBL, teachers, learners, and community members work together to construct shared understanding of the activity (Krajcik and Blumenfeld 2006). The use of collaborative groups, then, is inherent in a PBL classroom. Learners are able to develop a more linked conceptual understanding between new information and prior knowledge by sharing, challenging, and expanding on the principles and ideas of others (Blumenfeld et al. 1996). Refining one another's ideas and challenging one another's understandings also helps foster a community of learners. Learners learn firsthand what it means to function as part of a community.

Cognitive Tools

Cognitive tools—such as visual aids, graphic organizers, computer software, and manipulatives—can amplify and build on what learners learn. These tools can help learners understand a conceptual idea, expand the range of questions that learners can investigate, and offer learning experiences that might not otherwise be possible. In a PBL classroom, the teacher uses cognitive tools as scaffolds to support learning and to assess learners' understanding of key concepts.

The Six A's

An exemplary PBL unit ensures that both instruction and content enable learners to master core competencies. The Six A's (Markham, Larmer, and Ravitz 2003, p. 34) can be used as a guide for designing a math PBL unit:

- **Authenticity:** The project is situated in the real world—other professionals are tackling the same problem or question addressed by the project. In addition, the problem has meaning and relevance to learners, and there is an appropriate audience to view learners' products.

- **Academic Rigor:** The Driving Question is well-defined and tightly integrated to the content standards. The project also demands breadth and depth of both specific knowledge and central concepts. Learners develop habits that are indicative of efficient and effective problem solvers, such as questioning and posing problems, applying past knowledge to new situations, employing precision of language and thought, and maintaining persistence.

- **Applied Learning:** Learners use multiple high-performance work organization skills, such as working in teams, communicating ideas, applying new knowledge to the problem, and organizing and analyzing information. Learners are able to identify and apply the self-management skills needed to improve their group's performance.

- **Active Exploration:** Learners conduct field-based activities, such as interviewing experts, surveying groups, and exploring worksites. Learners gather information from various sources and use appropriate methods to obtain the needed data.

This piece can be challenging for mathematics educators. Because the Six A's criteria are not discipline-specific, teachers may overlook important criteria that are specific to a mathematics classroom. For example, while gathering information from a variety of sources and using a variety of methods to solve the problem, such as interviews, model building, and online research, are exemplary characteristics of Active Exploration (Markham, Larmer, and Ravitz 2003), teachers are often unsure how these might translate to a mathematics classroom. Chapter 13: Tips for Teachers from a PBL Mathematics Educator discusses in more detail how teachers can encourage their learners to actively explore in a mathematics classroom.

- **Adult Connections:** Learners are provided with mentorship opportunities, where they work alongside adults at a worksite relevant to the project. Learners develop meaningful relationships with members of the community who have expertise and experience in a particular field.

- **Assessments:** Various formal and informal assessments occur intermittently throughout the project, and learners are given timely feedback from both peers and teachers. The project requires multiple products, all of which are aligned with the project's ultimate goal. The project culminates in an exhibition or presentation for an informed audience.

Assessment

Principles to Actions: Ensuring Mathematical Success for All (NCTM 2014, pp. 91–92) describes several principles of assessments that support effective teaching and learning:

- Assessment is an ongoing process that is embedded in instruction to support student learning and to allow teachers to adjust their instruction as needed.

- Mathematical understanding and processes can be measured through the use of a variety of assessment strategies and tasks.

- Multiple data sources are needed to provide an accurate picture of both teacher and student performance.

- Assessment is a process that should help students become better judges of their own work, assist them in recognizing high-quality work when they produce it, and support them in using evidence to advance their own learning.

Meaningful assessments provide information on learners' mastery of key mathematical ideas and enable educators to draw conclusions about their own progress in achieving the learning goals of the unit. Rigorous math PBL units must include learners' mastery of mathematics ideas in the project rubric.

When designing and implementing a PBL unit, it is crucial that key mathematical ideas, learning objectives, instructional tasks and methods, and assessment practices all align. NCTM's *Curriculum and Evaluation Standards for School Mathematics* (1989) summarizes this issue clearly: "The degree to which meaningful inferences can be drawn from [assessments] depends on the degree to which the assessment methods and tasks are aligned or are in agreement with the curriculum" (p. 193).

References

Barab, Sasha A., and Thomas Duffy. "From Practice Fields to Communities of Practice." In *Theoretical Foundations of Learning Environments*, edited by David H. Jonassen and Susan M. Land, pp. 25–55. Mahwah, N.J.: Lawrence Erlbaum Associates, 2000.

Blumenfeld, Phyllis C., Ronald W. Marx, Joseph S. Krajcik, and Elliot Soloway. "Learning with Peers: From Small Group Cooperation to Collaborative Communities." *Educational Researcher* 25 (1996): 37–40.

Brown, John S., Allan Collins, and Paul Duguid. "Situated Cognition and the Culture of Learning." *Educational Researcher* 18 (1989): 32–42.

Dewey, John. *The Child and the Curriculum.* Chicago: University of Chicago Press, 1902.

Gallagher, Shelagh A., and William J. Stepien. "Content Acquisition in Problem-Based Learning: Depth versus Breadth in American Studies." *Journal for the Education of the Gifted* 19 (1996): 257–275.

Goldman, Susan R., Ted S. Hasselbring, and the Cognition and Technology Group at Vanderbilt. "Achieving Meaningful Mathematics Literacy for Students with Learning Disabilities: Challenges and Resources." *Journal of Learning Disabilities* 30 (1996): 198–208.

KnowledgeWorks Foundation. *Future Forces Affecting Education*, 2009. http://www.kwfdn.org.

Krajcik, Joseph S., and Phyllis Blumenfeld. "Project-Based Learning." In *The Cambridge Handbook of the Learning Sciences*, edited by R. Keith Sawyer, pp. 317–334. New York: Cambridge, 2006.

Larmer, John, David Ross, and John R. Mergendollar. *PBL Starter Kit: To-the-Point Advice, Tools and Tips for Your First Project in Middle or High School.* San Rafael, Calif.: Buck Institute for Education, 2009.

Markham, Thom, John Larmer, and Jason Ravitz. *Project Based Learning Handbook: A Guide to Standards-Focused Project Based Learning.* 2nd ed. Novato, Calif.: Buck Institute for Education, 2003.

National Council of Teachers of Mathematics (NCTM). *Curriculum and Evaluation Standards for School Mathematics.* Reston, Va.: Author, 1989.

National Council of Teachers of Mathematics (NCTM). *Focus in High School Mathematics: Reasoning and Sense Making.* Reston, Va.: Author, 2009.

National Council of Teachers of Mathematics (NCTM). *Principles to Actions: Ensuring Mathematical Success for All.* Reston, Va.: Author, 2014.

National Governors Association and Council of Chief State School Officers (NGA and CCSSO). *Common Core State Standards Initiative: Mathematics*, 2010. http://www.corestandards.org/the-standards/mathematics.

Partnership for 21st Century Learning. *21st Century Curriculum and Instruction*, n.d. http://www.p21.org.

Piaget, Jean. *Science of Education and the Psychology of the Child*. New York: Viking Press, 1970.

Plucker, Jonathan A., and Jeffery A. Nowak. "Creativity in Science for K–8 Practitioners: Problem-Based Approaches to Discovery and Invention." In *Teaching the Creative Child K–8*, edited by Mervin D. Lynch and Carol Ruth Harris, pp. 145–158. Boston: Allyn & Bacon, 2000.

Resnick, Lauren B. *Education and Learning to Think*. Washington, D.C.: The National Academies Press, 1987.

Stein, Mary Kay, Randi A. Engle, Margaret S. Smith, and Elizabeth K. Hughes. "Orchestrating Productive Mathematical Discussions: Five Practices for Helping Teachers Move Beyond Show and Tell." *Mathematical Thinking and Learning* 10 (2008): 313–340.

Vygotsky, Lev. S. *Mind in Society: The Development of Higher Psychological Processes*. Cambridge, Mass.: Harvard University Press, 1978.

Vygotsky, Lev S. *Thought and Language*. Cambridge, Mass.: MIT Press, 1962.

Template 1.1: Scaffolding NTKs

Scaffolding NTKs: Activity and Assessment Planning

Anticipated Knowledge and Skills Students Need (NTKs)	Assignment or Activity to Address NTKs	How Assignment or Activity Will Be Addressed	Learning Outcomes Addressed in Assignment/Activity

Template adapted from the Buck Institute for Education (www.bie.org).

Template 1.2: Project Planning Form

Name of Project:	
Designed by (Teacher Name[s] and Email Address[es]):	

Project Idea What is the issue, problem or theme?	
Topic(s) addressed: List one or more topics this project addresses.	
Essential Question What is the Driving Question or challenge?	
Entry Event What is the hook to launch this project?	

CCSSM and Standards for Mathematical Practices
List those to be addressed by the project.

T = Taught; P = Practiced; A = Assessed	T	P	A	T = Taught; P = Practiced; A = Assessed	T	P	A
Written communication	☐	☐	☐	Technology literacy	☐	☐	☐
Oral communication	☐	☐	☐	Work ethic	☐	☐	☐
Collaboration	☐	☐	☐	Civic responsibility	☐	☐	☐
Critical thinking	☐	☐	☐	Numeracy	☐	☐	☐
Information literacy	☐	☐	☐	Core content skills	☐	☐	☐

Learner Outcomes
Note the 21st-century skills taught, practiced, and/or assessed in this project

Habits of Mind:
Indicate one or two habits of mind that are the focus of this project.

- ☐ Persisting
- ☐ Managing impulsivity
- ☐ Listening to others
- ☐ Thinking flexibly
- ☐ Thinking about thinking
- ☐ Striving for accuracy/precision
- ☐ Questioning/posing problems
- ☐ Applying past knowledge
- ☐ Communicating with clarity
- ☐ Gathering data, using all senses
- ☐ Creating, imagining, innovating
- ☐ Responding with awe
- ☐ Taking responsible risks
- ☐ Finding humor
- ☐ Thinking interdependently
- ☐ Learning continuously

Student Production	Group products (major types):	
	Individual products (major types):	

Presentation Audience

Check all that apply:
- ☐ Class
- ☐ School
- ☐ Community
- ☐ Experts
- ☐ Web (public)
- ☐ Parents
- ☐ Other:

Template adapted from the Buck Institute for Education (www.bie.org).

Template 1.2: Project Planning Form (*continued*)

Assessments and Reflection	Rubric(s) Check and describe all that will be used for this project.	☐ Multimedia presentation rubric	☐ Other:
		☐ Oral presentation rubric	☐ Other:
		☐ CCSS ELA & literacy writing rubrics	☐ Other:
		☐ School writing rubric	☐ Other:
		☐ School learner outcomes rubric	☐ Other:
	Assessment type(s) Check and describe all that will be used for this project.	☐ Quiz:	☐ Performance assessment:
		☐ Test:	☐ Notes review:
		☐ Essay:	☐ Checklist:
		☐ Online assessment:	☐ Concept maps:
	Reflection tools Check and describe all that will be used for this project.	☐ Survey:	☐ Focus Group
		☐ Discussion:	☐ Personal learning plan
		☐ Journal	☐ Student-teacher conference
Project Resources	On-site personnel:		
	Technology:		
	Community resources:		
	Print resources:		
	Online resources:		

Template adapted from the Buck Institute for Education (www.bie.org).

Template 1.3: Project Calendar

Monday	Tuesday	Wednesday	Thursday	Friday
		WEEK 1		
		WEEK 2		
		WEEK 3		
		WEEK 4		

Template 1.4: Six Problem-Solving Phases Planning Worksheet

PBL Process	NTKs	Scaffolding	Before Moving to Next Phase...
Phase 1 What is the need or problem?			
Phase 2 What needs to be in our solution?			
Phase 3 What are possible solutions?			
Phase 4 What solution should we use?			
Phase 5 How do we create, run, and inspect our solution?			
Phase 6 Reflect back: Did we solve the problem? Do we have a new problem?			

Template adapted from the Buck Institute for Education (www.bie.org).

Fostering College and Career Readiness in PBL Classrooms

Enrique Galindo, *Indiana University Bloomington*
Jean Lee, *University of Indianapolis, Indiana*

The goal of precollege education is to prepare students for both college and careers. But while college and career readiness are sometimes seen as the same thing, they are in fact quite different. *College readiness* involves having the skills needed to enroll in and successfully complete postsecondary education. *Career readiness* includes having the skill set that allows one to enter true career pathways that offer family-sustaining wages and opportunities for advancement (Association for Career and Technical Education [ACTE] 2010). Two key elements to ensure that learners are both college and career ready are (1) a rigorous curriculum that is relevant to the students and (2) a learning environment where positive relationships nurture student learning. In this chapter, we outline how PBL can be a means to support college and career readiness, to make the mathematics classroom come alive, and to address rigor, relevance, and relationships in the classroom.

What Is Career Readiness?

The Career Readiness Partner Council (CRPC) (2012) provides the following definition:

> A career-ready person effectively navigates pathways that connect education and employment to achieve a fulfilling, financially-secure and successful career. A career is more than just a job. Career readiness has no defined endpoint. To be career ready in our ever-changing global economy requires adaptability and a commitment to lifelong learning, along with mastery of key knowledge, skills and dispositions that vary from one career to another and change over time as a person progresses along a developmental continuum. Knowledge, skills and dispositions are inter-dependent and mutually reinforcing. (p. 2)

ACTE describes career readiness as involving three major skill areas: core academic skills, employability skills, and technical skills related to a specific career pathway.

- **Core academic skills:** ACTE, Achieve, and other organizations have stated that career-ready core academics and college-ready core academics are essentially the same, thus creating overlap in the preparation students need to be ready for postsecondary education and careers. Academic skills are explicitly addressed in standards documents, such as the Common Core State Standards for Mathematics; however, in order to be career ready, students must be able to apply these academic skills in context and in authentic situations.

- **Employability skills** are the skills often cited by employers as most critical to workplace success in the 21st-century economy, including critical thinking, adaptability, problem solving, oral and written communication, collaboration and teamwork, creativity, responsibility, professionalism, ethics, and technology use (CRPC 2012). Too often these skills are not addressed in conjunction with the academic content.

- **Technical skills** include job-specific knowledge and skills.

Why Are College and Career Readiness Important?

In a recent study conducted by Hart Research Associates on behalf of the Association of American Colleges and Universities (AACU) (2015), 613 U.S. college students were asked to rate their perceived proficiency in skills necessary to succeed in today's workforce. Four hundred hiring employers were also asked to rate recent college graduates on those same skills. The results were surprising: College students' ratings on their perceived proficiency of critical workforce skills were higher than how employers rated them. For example, 59 percent of college graduates believe that they can solve complex problems in the workplace, while only 24 percent of employers say that they find this to be true. Figure 2.1 displays some of the largest discrepancies between student and employer responses (AACU 2015, p. 12).

These and similar findings illustrate a disconnect between the perceptions of recent college graduates and hiring managers.

The skills needed for the workforce of the future are not the same skills students needed a generation ago. In some cases, this means a "new normal" for college and career readiness, where students learn about postsecondary options earlier and engage in hands-on, real-world lessons. While proponents of career and technical education have reported a divide between teaching for college readiness vs. career readiness (Achieve 2012), PBL instruction embeds the use of soft skills (e.g., communicating effectively, thinking critically, working successfully on a team) *in conjunction with* mastery of the academic content. Students demonstrate use of these skills throughout the unit, thereby fostering their use in the real world.

The PBL units in this book demonstrate that some of these soft skills are encouraged, while others are explicitly taught and assessed. By incorporating these skills into instruction, the classwork becomes relevant to students whether they choose college, the workplace, the military, or a certification training program.

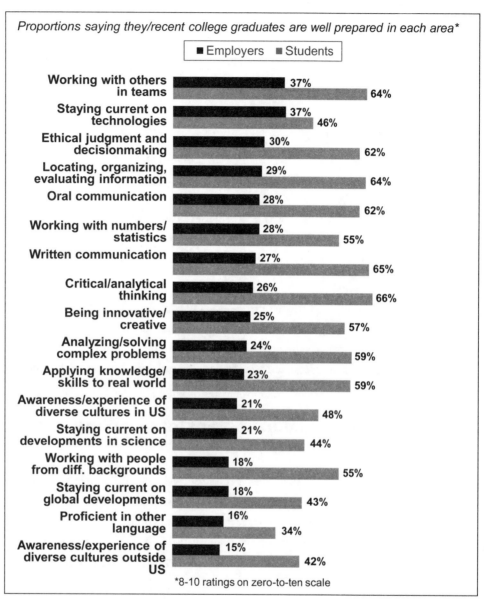

Proportions saying they/recent college graduates are well prepared in each area*

■ Employers   ■ Students

	Employers	Students
Working with others in teams	37%	64%
Staying current on technologies	37%	46%
Ethical judgment and decisionmaking	30%	62%
Locating, organizing, evaluating information	29%	64%
Oral communication	28%	62%
Working with numbers/ statistics	28%	55%
Written communication	27%	65%
Critical/analytical thinking	26%	66%
Being innovative/ creative	25%	57%
Analyzing/solving complex problems	24%	59%
Applying knowledge/ skills to real world	23%	59%
Awareness/experience of diverse cultures in US	21%	48%
Staying current on developments in science	21%	44%
Working with people from diff. backgrounds	18%	55%
Staying current on global developments	18%	43%
Proficient in other language	16%	34%
Awareness/experience of diverse cultures outside US	15%	42%

*8-10 ratings on zero-to-ten scale

Reprinted with permission from *Falling Short? College Learning and Career Success.* Copyright 2015 by the Association of American Colleges and Universities.

Fig. 2.1. Discrepancies between employers' and college graduates' scores re: preparedness across learning outcomes.

Common Core State Standards for Mathematics

Over the past two decades, various reform documents (e.g., National Council of Teachers of Mathematics [NCTM] 2000, 2014; National Governors Association and Council of Chief State School Officers [NGA/CCSSO] 2010) have emphasized the importance of students' understanding of mathematics content and the ways in which students engage in the learning of mathematics. These documents advocate for student engagement in authentic problem solving. In particular, *Principles and Standards for School Mathematics* (NCTM 2000) proposes that students must learn mathematics with understanding, actively building new knowledge from experience and prior knowledge.

The release of the CCSSM spurred much discussion about what it means to be *college and career ready* in mathematics. The CCSSM (NGA/CCSSO 2010) are intended to prepare students to be college and career ready by stating a set of expectations (i.e., knowledge, skills, and abilities) of what students will be able to demonstrate after completing a particular grade. In general, the standards seek to prepare students to think and reason mathematically, which is an essential skill for college and career readiness. The CCSSM include eight Standards for Mathematical Practice (SMP) for each grade that highlight the importance of students making sense of problems and persevering in solving them, reasoning abstractly and quantitatively, constructing viable arguments and critiquing the reasoning of others, modeling with mathematics, using tools, attending to precision, looking for and making use of structure, and looking for and expressing regularity in repeated reasoning. These standards call on students to apply mathematical ways of thinking to real-world issues and challenges.

The standards' emphasis on mathematical modeling and other mathematical practices has teachers wondering what these practices look like and how students learn them. PBL offers an effective way to address significant content in mathematics learning environments with an additional emphasis on how mathematics content and practices can be used to solve problems, answer questions, and address the challenges in industry, agencies, and other workplaces in the students' communities.

How Project-Based Learning Fosters College and Career Readiness

In the *Project Based Learning Handbook*, Markham, Larmer, and Ravitz (2003) define PBL as "a systematic teaching method that engages students in learning knowledge and skills through an extended inquiry process structured around complex, authentic questions and carefully designed projects and task" (p. 4). The Buck Institute for Education (n.d.) identifies eight essential elements of PBL (many of which we discussed in chapter 1) that help foster important college and career readiness skills:

- **Focus on significant content.** At its core, the project is focused on teaching students important knowledge and skills, derived from standards and key concepts at the heart of academic subjects.

- **Develop 21st-century competencies.** Students build competencies that are especially valuable for today's world, such as critical thinking, problem solving, collaboration, communication, and creativity/innovation, all of which are taught and assessed.

- **Engage students in in-depth inquiry.** Students are engaged in a rigorous, extended process of asking questions, using resources, and developing answers.

- **Organize tasks around a driving question.** Project work is focused by an open-ended question that students understand and find intriguing, which captures their task or frames their exploration.

- **Establish NTKs.** Students see the need to gain knowledge, understand concepts, and apply skills in order to answer the Driving Question and create a product, beginning with an Entry Event that generates their interest and curiosity.

- **Encourage voice and choice.** Students are allowed to make some choices about the products to be created, how students work, and how they use their time, guided by the teacher and depending on their age level and PBL experience.
- **Incorporate critique and revision.** The project includes processes for students to give and receive feedback on the quality of their work, leading them to make revisions or conduct further inquiry as needed.
- **Include a public audience.** Students present their work to people in the community, not just their classmates and teacher. This raises both the level of discourse and the authenticity of the project, motivating students even further because they feel validated and heard.

Rigor, Relevance, and Relationships

Educators working with PBL use the criteria of *rigor*, *relevance*, and *relationship*s as a guideline for successful learning environments.

Rigor is related to at least three of the Six A's described in chapter 1: Academic Rigor, Applied Learning, and Assessment Practices. PBL provides opportunities for students to deeply engage with important content knowledge, apply that knowledge to tackle challenging situations, and demonstrate their understanding by finding an answer to the Driving Question and in the project assessments.

Relevance is related to at least two of the Six A's: Authenticity and Applied Learning. By working on projects that are situated in the real world, presenting products to an authentic audience, and using knowledge to solve challenges related to current or historical events or their personal backgrounds, students are more likely to find learning meaningful and relevant.

Authentic audience: Stakeholders in the students' project (e.g., community members, experts in the topic, the principal) who provide a real-world connection to the students' work in the classroom.

The third criterion, **relationships**, is related to at least two of the Six A's: Active Exploration and Adult Connections. Well-designed PBL projects encourage students to work effectively with their peers and with other adults (Kubik 2013).

These elements are cross-cutting. For example, it is difficult to achieve rigor without designing a project that also fosters positive relationships. Supporting students in understanding themselves as learners helps them approach projects with the confidence of knowing the assets they can bring to their team, which in turn results in productive collaborations with team members. Once students get comfortable with the fact that different team members bring different skills and knowledge to the project work, they can also see the value of working with adults other than their teacher. Since their teacher cannot be an expert on all subject matter, collaboration with other adults who have expertise on the topic of the project contributes to a quality product.

The PBL units in this book are classroom-tested projects that provide opportunities for students to learn the rigorous mathematics necessary for college and career readiness; engage students in authentic and relevant problems, challenges, and issues; and foster positive relationships among the school, teacher, students, and community members. Students learn mathematics with conceptual understanding and procedural fluency while actively engaging in authentic problem solving.

References

Achieve. *Common Core State Standards and Career and Technical Education: Bridging the Divide between College and Career Readiness*, 2012, http://www.achieve.org/files/CCSS-CTE-BridgingtheDivide.pdf.

Association for Career and Technical Education (ACTE). *What Is "Career Ready"?*, 2010, https://www.acteonline.org/WorkArea/DownloadAsset.aspx?id=2114.

Association of American Colleges and Universities. *Falling Short? College Learning and Career Success*, 2015, http://www.aacu.org/leap/public-opinion-research/2015-survey-results.

Buck Institute for Education (BIE). *Why Project Based Learning (PBL)?*, n.d., http://www.bie.org.

Career Readiness Partner Council (CRPC). *Building Blocks for Change: What It Means to Be Career Ready*, 2012, http://www.careerreadynow.org/docs/CRPC_4pager.pdf.

Kubik, Tim. "Quality Projects: Put Relationships before the Rigor." PBL Blog, October 14, 2013, http://bie.org/blog/quality_projects_put_relationships_before_the_rigor.

Markham, Thom, John Larmer, and Jason Ravitz. *Project Based Learning Handbook: A Guide to Standards-Focused Project Based Learning*. 2nd ed. Novato, Calif.: Buck Institute for Education, 2003.

National Council of Teachers of Mathematics (NCTM). *Principles and Standards for School Mathematics*. Reston, Va.: Author, 2000.

National Council of Teachers of Mathematics (NCTM). *Principles to Actions: Ensuring Mathematical Success for All*. Reston, Va: Author, 2014.

National Governors Association and Council of Chief State School Officers (NGA and CCSSO). *Common Core State Standards Initiative: Mathematics*, 2010, http://www.corestandards.org/the-standards/mathematics.

Benefits of PBL Instruction: A Research Overview

Jill Bradley-Levine, *Ball State University, Muncie, Indiana*

Gina G. Romano, *Indiana University Bloomington*

Effective learning is linked to opportunities to "explore, inquire, solve problems, and think critically" (Asghar et al. 2012, p. 86). Toward this end, concentrated reform initiatives across many content areas have integrated authentic and student-driven instructional approaches. Although these initiatives have different names (including inquiry learning, deeper learning, expeditionary learning, problem-based learning, and PBL), they share the common goal of engaging students through exploring real-world issues and solving practical problems.

[handwritten note: where is the research that supports PBL for middle school math?]

Much of the research on PBL has focused on student outcomes in the secondary or postsecondary context, especially in the areas of science, technology, engineering, and mathematics (STEM). For example, postsecondary-level studies have examined engineering and medicine (Bédard et al. 2012; Krishnan, Gabb, and Vale 2011; Sockalingam and Schmidt 2011). High school-level studies have looked at such topics as marine engineering and physical science (Verma, Dickerson, and McKinney 2011) and general STEM knowledge and interest (Lou et al. 2011). A few studies have also focused on middle and elementary school science (Eskrootchi and Oskrochi 2010; Krajcik et al. 1998; Panasan and Nuangchalerm 2010).

PBL requires students to tackle complex challenges, questions, and problems through the development of products and presentations (Buck Institute for Education [BIE] 2012; Thomas 2000). The PBL process provides students with opportunities to develop their own questions to drive learning, to study concepts and information to answer those questions, and to apply that knowledge to products they develop (BIE 2012). PBL encourages rigorous learning because it requires students to take an active role in understanding concepts and content related to projects, and it enables them to develop 21st-century skills that will increase their workforce readiness. Since students are able to apply classroom content to real-life phenomena, PBL also facilitates career exploration, technology use, student engagement, community connections, and content relevancy (Blumenfeld et al. 1991; BIE 2012).

Effects of PBL on Student Outcomes

Multiple studies have reported that students in PBL-taught classrooms demonstrate improved critical thinking and problem-solving skills (Boaler 1997; Finkelstein et al. 2010; Penuel and Means 2000; Shepherd 1998; Tretten and Zachariou 1995). Researchers have also found that PBL is a successful way of teaching 21st-century skills, such as collaboration, and that it improves student engagement and content learning (Barron et al. 1998; Belland, Ertmer, and Simons 2006; Boaler 1997; Brush and Saye 2008; ChanLin 2008; Finkelstein et al. 2010; Horan, Lavaroni, and Beldon 1996; Lightner, Bober, and Willi 2007; Mioduser and Betzer 2003; Ravitz and Mergendoller 2005; Tretten and Zachariou 1995; Verma, Dickerson, and McKinney 2011). Further, students demonstrate greater initiative by seeking and using resources and revising their work, behaviors that were uncharacteristic of them before they were immersed in PBL-instructed classes (Barron et al. 1998).

Strobel and van Barneveld (2009) conducted a qualitative meta-synthesis of meta-analyses to identify generalizable findings regarding the effectiveness of PBL in teaching content knowledge. They conclude that traditional instruction produces better outcomes when assessing basic knowledge, but that PBL produces better results when assessing clinical knowledge and skills and retention of those skills:

> PBL is significantly more effective than traditional instruction to train competent and skilled practitioners and to promote long-term retention of knowledge and skills. (p. 55)

[handwritten margin note: What kind of "traditional" instruction are they referring to?]

PBL has been found to have positive effects on four specific student outcomes—higher-order thinking skills, collaboration skills, student engagement, and student content learning—each of which is described in more detail below.

Higher-Order Thinking Skills

Several studies have found that PBL has a positive effect on the development of higher-order thinking skills, such as critical thinking and problem solving, for specific groups of students. In particular, students with average to low verbal ability and students with little previous content knowledge demonstrated increased learning in PBL-taught classes compared to students in traditionally taught classes (Mergendoller, Maxwell, and Bellisimo 2006). Another study demonstrated that PBL positively impacts low-ability students, who exhibited critical thinking skills (such as synthesizing, evaluating, predicting, and reflecting) as a result of being immersed in PBL-taught classes (Horan, Lavaroni, and Beldon 1996). It is worth noting that high-ability students also improved their use of these skills (Horan, Lavaroni, and Beldon 1996).

Collaboration Skills

Research has shown that PBL fosters collaboration skills in a variety of students. After working in groups in PBL-taught classes, elementary school students learned to understand things from multiple perspectives and to demonstrate conflict resolution skills (ChanLin 2008). Special education students developed social skills, such as patience and empathy (Belland, Ertmer, and Simons 2006), and low-ability students demonstrated initiative, management, teamwork, and conscientiousness (Horan,

Lavaroni, and Beldon 1996). Students also reported that they enjoyed PBL work because it gave them opportunities to interact with their friends and to make new friends through cooperative projects (Belland, Ertmer, and Simons 2006; Lightner, Bober, and Willi 2007).

Student Engagement

Researchers have found high levels of student engagement in PBL classrooms (Belland, Ertmer, and Simons 2006; Brush and Saye 2008) due to PBL's real-world, problem-solving context (Blumenfeld et al. 1991). A study of one economics class revealed that a PBL unit successfully engaged both the lowest- and highest-performing students, as well as students who were least interested in economics at the start of the unit (Ravitz and Mergendoller 2005). Another study reported that PBL had a positive effect on students' motivation to learn: Elementary school teachers who reported devoting 37 percent of their overall instruction time to PBL said that students' work ethic improved, as well as their confidence and attitudes toward learning (Tretten and Zachariou 1995). Similarly, when secondary school students from high-need schools participated in an applied shipbuilding project, they were more enthusiastic about marine engineering and physical science (Verma, Dickerson, and McKinney 2011).

Student Content Learning

Those immersed in PBL-taught classes emerge with more useful, real-world content knowledge that can be applied to a variety of tasks (Boaler 1997). An experimental study of 76 teachers who used PBL revealed that compared to the control group of students in traditionally taught classes, their students scored higher both on standardized exams and on ability tests that measured problem-solving skills and content application to real-world problems (Finkelstein et al. 2010). In addition, one study found that students were able to demonstrate specific content-area skills after taking part in a PBL unit (Mioduser and Betzer 2003). For example, among students using measurement skills to develop blueprints for a geometry project involving architecture and design, 84 percent designed architectural plans for a playhouse that faculty judged could actually be built (Barron et al. 1998).

Groups Underrepresented in STEM

Researchers have studied students' interest in STEM careers after they experienced PBL instruction, looking in particular at students who are traditionally underrepresented in STEM careers, such as girls and women and students from low socioeconomic backgrounds (Lou et al. 2011; Verma, Dickerson, and McKinney 2011). After examining high school marine engineering and physical science courses taught using the PBL approach, Verma and colleagues (2011) found that students traditionally underrepresented in STEM reported higher levels of engagement and interest in STEM subjects after participating in PBL. Lou and colleagues (2011) also found that girls and women demonstrated greater interest in STEM after experiencing PBL. Eskrootchi and Oskrochi (2010) found that both girls and those students who might not learn effectively through more traditional instructional approaches can succeed from new technological innovations combined with PBL in STEM areas such as computer technology.

Eskrootchi and Oskrochi's (2010) quasi-experimental study of PBL instruction in a technology-rich environment determined the combined impact of technology integration and PBL on students. Eighth graders were divided into three groups: a control group who received traditional, lecture-based instruction; an experimental group who performed a simulation model using only technology; and an experimental group who learned through PBL while also using technology. Students were tested on both conceptual and content knowledge. Students in the experimental group that learned through PBL while using technology outperformed the other two groups in subject understanding but not content knowledge. In other words, although students in all three groups scored similarly in content knowledge, students who learned via PBL more fully grasped the abstract foundations of the subject, which demonstrates deeper learning than merely having basic knowledge of the subject. Additionally, the PBL project had a statistically significantly stronger effect on girls, who earned higher mean scores in the experimental group that learned through PBL while using technology than girls in the control group.

This demonstrates that effectively implementing technology with PBL results in improved student achievement compared to students using technology alone. The researchers believe that this is due to increased student collaboration, authenticity, and the establishment of spaces for more equitable contribution, all of which are present in PBL-instructed classes.

Implementing Instruction

Many teachers perceive that PBL was beneficial to their students, thus motivating them to adopt this instructional approach for their classes. A national survey of public school teachers revealed that they were likely to use PBL because they believe it teaches abilities beyond academic content, including such 21st century skills as collaboration and presentation techniques (Ravitz 2008). In addition, after interviewing and observing 10 sixth-grade science teachers implementing technology-supplemented PBL, Liu and colleagues (2012) found that teachers use PBL if they believe that it addresses content standards, aligns with their philosophy of teaching, provides an innovative form of instruction that fosters 21st-century skills, challenges students in an engaging way that meets diverse learning needs, and is supported by building administrators.

Conclusion

PBL has been found to be a highly effective instructional approach for teaching and learning 21st-century skills, and has the potential to broaden all students' interest in STEM fields. To that end, nonprofit organizations and industry leaders have embraced PBL as a means of addressing the need for an educated and well-prepared workforce, particularly in STEM areas. According to Robert Abrams of Economic Opportunities through Education by 2015:

> Building core skills in STEM curricula is a key element of our workforce development strategy, and we have found that PBL is a very effective model for engaging students, enabling them to succeed academically, and helping them develop proficiency in the workplace skills highly valued by employers. (Personal communication, January 22, 2014)

For these reasons, K–12 math teachers should explore ways to use the PBL instructional approach with their students.

References

Asghar, Anila, Roni Ellington, Eric Rice, Francine Johnson, and Glenda M. Prime. "Supporting STEM Education in Secondary Science Contexts." *Interdisciplinary Journal of Problem-Based Learning* 6 (2012): 85–125.

Barron, Brigid, Daniel Swartz, Nancy Vye, Allison Moore, Anthony Petrosino, Linda Zech, John Bransford, and Cognition and Technology Group at Vanderbilt. "Doing with Understanding: Lessons from Research on Problem- and Project-Based Learning." *The Journal of the Learning Sciences* 7 (1998): 271–311.

Bédard, Denis, Christelle Lison, Daniel Dalle, D. Côté, and Noël Boutin. "Problem-Based and Project-Based Learning in Engineering and Medicine: Determinants of Students' Engagement and Persistence." *Interdisciplinary Journal of Problem-Based Learning* 6 (2012): 7–30.

Belland, Brian R., Peggy A. Ertmer, and Krista D. Simons. "Perceptions of the Value of Problem-Based Learning among Students with Special Needs and Their Teachers." *Interdisciplinary Journal of Problem-Based Learning* 1 (2006): 1–18.

Blumenfeld, Phyllis C., Elliot Soloway, Ronald W. Marx, Joseph S. Krajcik, Mark Guzdial, and Annemarie Palincsar. "Motivating Project-Based Learning: Sustaining the Doing, Supporting the Learning." *Educational Psychologist* 26 (1991): 369–398.

Boaler, Jo. *Experiencing School Mathematics: Teaching Styles, Sex, and Settings.* Buckingham, UK: Open University Press, 1997.

Brush, Thomas, and John Saye. "The Effects of Multimedia-Supported Problem-Based Inquiry on Student Engagement, Empathy, and Assumptions about History." *Interdisciplinary Journal of Problem-Based Learning* 2 (2008): 21–56.

Buck Institute for Education (BIE). *What Is PBL?*, 2012, http://www.bie.org/about/what_is_pbl.

ChanLin, Lih-Juan. "Technology Integration Applied to Project-Based Learning in Science." *Innovations in Education and Teaching International* 45 (2008): 55–65.

Eskrootchi, Rogheyeh, and G. Reza Oskrochi. "A Study of the Efficacy of Project-Based Learning Integrated with Computer-Based Simulation—STELLA." *Educational Technology & Society* 13 (2010): 236–245.

Finkelstein, Neal D., Thomas Hanson, Chun-Wei Huang, Becca Hirschman, and Min Huang. *Effects of Problem Based Economics on High School Economics Instruction.* (NCEE 2010-4002). Washington, D.C.: National Center for Education Evaluation and Regional Assistance, Institute of Education Sciences, U.S. Department of Education, 2010.

Horan, Carolyn, Charles Lavaroni, and P. Beldon. *Observation of the Tinker Tech Program Students for Critical Thinking and Social Participation Behaviors.* Novato, Calif.: Buck Institute for Education, 1996.

Krajcik, Joseph, Phyllis C. Blumenfeld, Ronald W. Marx, Kristin M. Bass, Jennifer Fredricks, and Elliot Soloway. "Inquiry in Project-Based Science Classrooms: Initial Attempts by Middle School Students." *The Journal of the Learning Sciences* 7 (1998): 313–350.

Krishnan, Siva, Roger Gabb, and Colleen Vale. "Learning Cultures of Problem-Based Learning Teams." *Australasian Journal of Engineering Education* 17 (2011): 67–78.

Lightner, Sharon, Marcie J. Bober, and Caroline Willi. "Team-Based Activities to Promote Engaged Learning." *College Teaching* 55 (2007): 5–18.

Liu, Min, Jennifer Wivagg, Renata Geurtz, Shih-Ting Lee, and Hyeseung Maria Chang. "Examining How Middle School Science Teachers Implement a Multimedia-Enriched Problem-Based Learning Environment." *Interdisciplinary Journal of Problem-Based Learning* 6 (2012): 46–84.

Lou, Shi-Jer, Ru-Chu Shih, C. Ray Diez, and Kuo-Hung Tseng. "The Impact of Problem-Based Learning Strategies on STEM Knowledge Integration and Attitudes: An Exploratory Study among Female Taiwanese Senior High School Students." *International Journal of Technology and Design Education* 21 (2011): 195–215.

Mergendoller, John R., Nan L. Maxwell, and Yolanda Bellisimo. "The Effectiveness of Problem-Based Instruction: A Comparative Study of Instructional Methods and Student Characteristics." *Interdisciplinary Journal of Problem-Based Learning* 1 (2006): 49–69.

Mioduser, David, and Nadav Betzer. "The Contribution of Project-Based Learning to High-Achievers' Acquisition of Technological Knowledge and Skills." *International Journal of Technology and Design Education* 18 (2008): 59–77.

Panasan, Mookdaporn, and Prasart Nuangchalerm. "Learning Outcomes of Project-Based Learning Activities." *Journal of Social Sciences* 6 (2010): 252–255.

Penuel, William R., and Barbara Means. "Designing a Performance Assessment to Measure Students' Communication Skills in Multimedia-Supported, Project-Based Learning." Paper presented at the Annual Meeting of the American Educational Research Association, New Orleans, La., April 2000.

Ravitz, Jason. "Project Based Learning as a Catalyst in Reforming High Schools." Paper presented at the Annual Meeting of the American Education Research Association, New York, N.Y., March 2008.

Ravitz, Jason, and John Mergendoller. "Evaluating Implementation and Impacts of Problem-Based Economics in U.S. High Schools." Paper presented at the Annual Meeting of the American Educational Research Association, Montreal, Canada, April 2005.

Shepherd, Norman Glenn. *The Probe Method: A Problem-Based Learning Model's Effect on Critical Thinking Skills of Fourth- and Fifth-Grade Social Studies Students.* Ph.D. Thesis, North Carolina State University, 1998.

Sockalingam, Nachamma, and Henk G. Schmidt. "Characteristics of Problems for Problem-Based Learning: The Students' Perspective." *Interdisciplinary Journal of Problem-Based Learning* 5 (2011): 6–33.

Strobel, Johannes, and Angela van Barneveld. "When Is PBL More Effective? A Meta-Synthesis of Meta-Analyses Comparing PBL to Conventional Classrooms." *Interdisciplinary Journal of Problem-Based Learning* 3 (2009): 44–58.

Thomas, John. "A Review of Research on Project-Based Learning." Report prepared for The Autodesk Foundation, 2000, http://bie.org/object/document/a_review_of_research_on_project_based_learning.

Tretten, Rudie, and Peter Zachariou. *Learning about Project-Based Learning: Assessment of Project-Based Learning in Tinkertech Schools.* San Rafael, Calif.: The Autodesk Foundation, 1995.

Verma, Alok K., Daniel L. Dickerson, and Sue McKinney. "Engaging Students in STEM Careers with Project-Based Learning—MarineTech Project." *Technology and Engineering Teacher* (September 2011): 25–31.

Solar Cooking with Conics

high school– advanced algebra?

Jacob Alan Goodman, *Ben Davis University High School, Indianapolis, Indiana*
Jean Lee, *University of Indianapolis, Indiana*

This three-week project was designed for a dual-credit college algebra course taught in an urban public high school where most students graduate with a two-year associate's degree. Many are first-generation college students. PBL was a districtwide initiative, and administrators were very supportive. Classes were held every other day for 90 minutes.

Solar Cooking with Conics addresses the Common Core algebra and geometry standards on conic sections by engaging students in the process of building a functional parabolic solar cooker while learning about local and global sustainability. Students are tasked with the challenge of building a solar cooker and using the sun's rays as the only means to heat soup.

The teacher had several goals for this unit in terms of students' college and career readiness:

- More than anything, the teacher wanted to increase students' exposure to a real-world issue. This project required students to use math in a way that millions of people around the world use it by having them build a solar cooker within a very small budget. Students started to internalize that having a safe way to heat water and food is a luxury rather than a guarantee for many people around the world—and that using math (i.e., designing a model within a budget) can literally save lives.

- The teacher also wanted to expose students to various ways of learning, combining new technologies with classic styles of learning, and having students interpret technical texts. The students collaborated digitally through Google applications. They constructed ellipses with thumbtacks and string. They broke down college-level textbooks, diagram by diagram, every day, in order to decode information that would help them design their solar cooker. Some students took on roles as heads of teams: a design team, a construction team, and a safety team. Students worked together like professionals, learned like researchers, and collaborated in savvy and productive ways.

Standards Addressed

This unit's focus on expressing geometric properties with equations—translating between the geometric description and the equation for a conic section—addressed three CCSSM for high school geometry:

- HSG.GPE.A.1: Derive the equation of a circle of given center and radius using the Pythagorean theorem; complete the square to find the center and radius of a circle given by an equation.
- HSG.GPE.A.2: Derive the equation of a parabola given a focus and directrix.
- HSG.GPE.A.3: Derive the equations of ellipses and hyperbolas given the foci, using the fact that the sum or difference of the distances from the foci is constant.

Other standards are also emphasized in the unit, including visualizing relationships between two-dimensional and three-dimensional (3-D) objects (HSG.GMD.B.4), applying geometric concepts in modeling situations (HSG.MG.A.1 and HSG.MG.A.3), writing expressions in equivalent forms to solve problems (HSA.SSE.B.3), interpreting functions that arise in applications in terms of the context (HSF.IF.B.4), and analyzing functions using different representations (HSF.IF.C).

In addition, this unit highlighted four SMPs:

- **Making sense of problems and persevering in solving them** (MP1) was an integral practice: Students needed to build a functional solar cooker by researching and designing the properties of a solar cooker and applying their knowledge about conic sections. When they realized that their first design was too complicated, they needed to go back and revise their plans.
- Students **reasoned abstractly and quantitatively** (MP2) by understanding the purpose of a focal point and its relationship to a parabola and understanding how to locate the focal point on their solar cooker.
- Students **used appropriate tools strategically** (MP5), both physical and electronic. They searched for different solar cooker designs online, investigated different conic sections using GeoGebra (interactive software that can be used for geometry, algebra, and calculus—see www.GeoGebra.org), used their textbook as a resource and reference, and learned how to safely use a box cutter.
- Students **looked for and expressed regularity in repeated reasoning** (MP8) by applying focal points in different contexts. For example, students investigated problems that involved locating the focal point in the headlight of a car so that it could reflect maximum brightness (see Day 5: Headlamp, below). They modeled the elliptical orbit of a planet's rotation around the sun and then transferred their reasoning to understanding how a sun's rays are collected at a focal point to heat a targeted area.

Project Highlights

The following lessons highlight how the *Solar Cooking with Conics* unit is rigorous, is relevant to students, and helped students build relationships with one another.

Details on specific lessons, activities, quizzes, and other resources can be found on the NCTM website (www.nctm.org/more4u).

Days 1 and 2: Launching the Project

The Entry Event for this project was a letter written by Nate Roberts, general manager of Pogue's Run Cooperative Grocer, a community-owned local grocery store. (See Sample Template 4.1.) In his letter, Nate Roberts pressed students to consider ways to use Earth's resources more thoughtfully.

The teacher showed students the project rubric (see Sample Template 4.2) and discussed the expectations for the project. Students chose a project manager for the entire class and then got into groups of four. Each small-group member then chose a role: building manager, safety manager, communication manager, or design manager. The class watched *Solar Cooking*, a National Geographic (n.d.) video that illustrated how people around the world struggle to purify water and cook food.

The teacher concluded the class by asking students to reflect on the following questions in their project journal:

1. For how many people in the world do you think solar cooking could be a life-saving technique?
2. Why would people use solar cookers? List three reasons.
3. Who will benefit most from the use of solar cookers?
4. What impact will solar cookers have on the environment?

Together, these activities garnered an enthusiastic reaction from the students. One student commented, "[Slang] got real today in math class!" before she started on the reflection prompts.

Talking in depth about solar cookers on the first day prepped the class for their work on the second day, during which they took apart 3-D conics manipulatives and interacted with Conic Section Explorer, an NCTM applet (http://illuminations.nctm.org/Activity.aspx?id=3506).

Day 5: Headlamp

Students were given a Challenge Packet of problems that they were each required to complete throughout the duration of the project. The problems were not necessarily harder than the class content, but they delved into learning the content at a deeper level.

For example, here is one problem in the Challenge Packet:

> A headlight is being constructed in the shape of a paraboloid with depth 4 inches and diameter 5 inches. Sketch a diagram of the headlamp. Then determine the distance d that the bulb should be from the vertex in order to have the beam of light shine straight ahead.

more**4**U

You can find additional Challenge problems on the NCTM website (www.nctm.org/more4u).

The teacher used GeoGebra to model this problem with sliders. Using such dynamic software helped students visualize the importance of the focal length and concavity of the parabola. Figure 4.1 shows a screen shot of the GeoGebra file the teacher created.

more**4**U

An investigation on the focal length of a solar cooker using GeoGebra is available on the NCTM website (www.nctm.org/more4u). Note that GeoGebra software is required in order to view these documents.

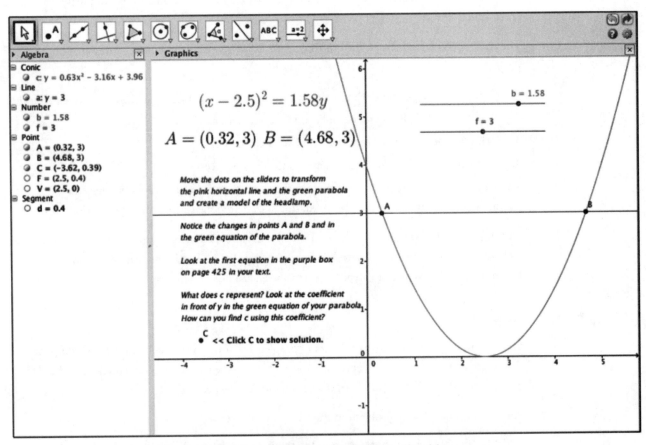

Fig. 4.1. Using GeoGebra for headlamp problem.

Relationship: Teach a few of the most tech-savvy students how to use GeoGebra in a small group first so they can help teach the rest of the students.

It is important to provide construction and investigative activities in which students explore the properties of conic sections. If teachers do not have the technology to use GeoGebra, they can construct an ellipse with string and paper for students to investigate foci. Allowing students to see this construction is meaningful because it reinforces the importance of the foci.

Day 7: United Nations Article

Students read an article from the United Nations Secretary-General's High-Level Panel on Global Sustainability (2012), which addressed such issues as lobbying, bi-corporations, worldwide civil unrest, depletion of natural resources, and consumerism. Students then reflected on the following questions:

- What does the report mean when it says, "Today our planet and our world are experiencing the best of times, and the worst of times?"

- How do you interpret, "We are not passive, helpless victims of the impersonal, determinist forces of history?" Look up the word *determinist*.

- Do you agree or disagree with the panel's vision? Support your opinion.

The teacher chose to have students read this article halfway through the project because he wanted to reinvigorate their sense of the importance of building a solar cooker and to recharge their interest.

Relevance: It is easy to get caught up in the content of the project, but it's important not to lose sight of the *context*—this is what drives students to do the work! Find ways to pump up the students mid-project by reminding them of the *why* behind their charge or challenge.

Relevance: Use authentic texts that talk about things the students really care about. Break the texts into smaller chunks, and stop often so everyone understands.

The teacher also noticed that incorporating literacy into his math classroom was improving students' writing over time. Having students frequently write reflections and responses to open-ended prompts (see fig. 4.2) throughout the unit dramatically increased their ability to critically reflect on the readings and videos.

---

1. How might Challenge Packet #3 relate to solar cooking? Be as detailed as possible. What knowledge and vocabulary did you gain in solving the problem?

2. Name one application of a conic section that interests you. Research that application, and share your discoveries.

3. What is the environmental impact of producing food locally versus shipping it thousands of miles from where it is grown (the way most food is produced now)? Research this question, and share what you learn.

Fig. 4.2. Sample open-ended prompts.

Day 15: Celebration

By design, the last day of the project coincided with Earth Day. The students organized and set up the celebration event; they invited their parents, guardians, and other teachers and students; and they designed a flyer to advertise their event (see fig. 4.3).

Event flyer designed by student

Fig. 4.3. Student-designed flyer advertising the celebration.

The teacher also invited both Nate Roberts, the community partner who helped launch this PBL unit, and the local news magazine to attend the celebration.

Community partner: A member of the community with an authentic connection to the project content who takes an active role in the PBL unit. Community partners bring new perspectives into the classroom and can serve as guest speakers, panel members for presentations, etc.

The teacher had several reasons for making this event public. He wanted his students to have a chance to celebrate all the work they had done. He also wanted to give them practice in using the skill of professionally presenting what they had learned to other people, including strangers, using academic language.

Relevance: Having students' products open to public scrutiny by people from the community lends another degree of authenticity to the project.

Relationship: Encourage students early on to invite their favorite adults—parents, friends, teachers, the principal, etc.—to the unit-end celebration of all they have learned.

On the last day of the project, students, teachers, parents, and community members came together to celebrate Earth Day and to test the parabolic solar cooker's ability to heat soup (see fig. 4.4). As the solar cooker began heating the soup (donated by Pogue's Run Cooperative Grocer), students gave their presentations to the invited guests.

Fig. 4.4. The functional student-built solar cooker.

The solar cooker worked! Students were able to heat the soup to well above the 165° goal. Extra soup was also available in crockpots, so everyone could enjoy soup and bread outdoors on this brisk, sunny day. It was clear that the students felt proud of their work.

References

National Geographic. *Solar Cooking*, n.d., http://video.nationalgeographic.com/video/solar-cooking.

United Nations Secretary-General's High-Level Panel on Global Sustainability. *Resilient People, Resilient Planet: A Future Worth Choosing—Overview.* New York: United Nations, 2012. https://sustainabledevelopment.un.org/index.php?page=view&nr=374&type=400&menu=35.

Sample Template 4.1. *Solar Cooking with Conics* Entry Event

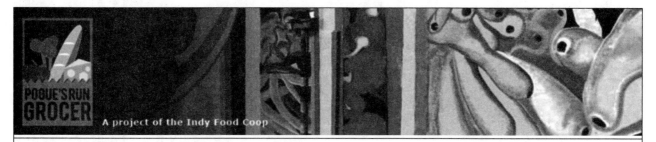

March 7, 2013

Dear Ben Davis University Scholars,

The world isn't getting any bigger. But the human population is, and with more people come more needs and a greater strain on our planet. We already can't feed all 7 billion of us, and the United Nations now predicts that the world population will grow to 10 billion by 2100.

Besides food, the energy needs required to maintain the lives of 10 billion people far exceeds what we are currently able to produce. In fact, if everyone in the world consumed as much energy as Americans, we would already be in an energy crisis. How will we become a sustainable species? What sources of food and energy are we currently not taking advantage of?

Pogue's Run Cooperative Grocer is a community-owned business invested in creating a local food economy and being a model for sustainable business. We have no owner or CEO. Our profits are reinvested in our business or distributed equally to all of our members. We would like you to use the properties of conic sections to create a solar cooker that can be used to harness the power of the sun to heat food. The cooker will be displayed at the University of Indianapolis and at Pogue's Run.

Your teacher, Mr. Goodman, has agreed to supervise this project, but it's up to you scholars to learn about conics and design and build the cooker. You will test the cooker at an event at your school. I will be there to make sure the cooker is safe and well-built. We will donate soup for your event, and you will serve that soup to your families and faculty at school. Make sure to keep a project journal so you can keep track of your progress and show your teachers and me all that you've learned. I suggest that you research local and global sustainability issues and the use of solar cookers around the world.

I'm sure you feel like there's a lot you need to know to complete this project. Be resourceful and consult with Mr. Goodman. I'm really looking forward to seeing what you come up with!

Sincerely,

Nate Roberts
General Manager
Pogue's Run Cooperative Grocer
2828 E. 10th Street
Indianapolis, IN 46201
317-426-4963

Sample Template 4.2: Solar Cooking With Conics Rubric

CRITERIA	Unsustainable (Barely Passing - C)	Sustainability Student (Acceptable - B)	Sustainability Scientist (Excellent - A)
Content Mastery (40%)	• Shows evidence of broad understanding of the geometric description and/or the equation for most conic sections (circle, parabola, ellipse, and hyperbola) • Developing visualization of relationships between the cone and some of its two-dimensional cross-sections. • Partially applies geometry in modeling the solar cooker. • Some evidence of interpreting quadratic functions that arise in applications. • Writes quadratic expressions in equivalent form. • Analyzes functions using at least one representation. • Earns between 70%–79% on final quiz	• Shows evidence of understanding how to translate between the geometric description and the equation for most conic sections (circle, parabola, ellipse, and hyperbola) • Shows evidence of visualizing relationships between the cone and some of its two-dimensional cross-sections. • Applies geometry in modeling the solar cooker. • Interprets quadratic functions that arise in the context of creating the solar cooker. • Writes quadratic expressions in equivalent form to solve problems. • Analyzes functions using at least two representations. • Earns between 80% and 89% on final quiz	• Shows evidence of deep understanding of how to translate between the geometric description and the equation for all conic sections (circle, parabola, ellipse, and hyperbola). • Shows evidence of visualizing relationships between the cone and its two-dimensional cross-sections. • Applies geometry in modeling the solar cooker in an efficient and sophisticated way. • Interprets quadratic functions that arise in applications in terms of the context. • Analyzes functions using multiple representations. • Earns 90% or higher on final quiz
Project Journal (30%)	• Submits incomplete journal: reflections, evidence of exploration, research, etc. missing (base requirements checklist will be provided) • Journal is unorganized or messy • Brings journal to class irregularly	• Submits completed journal with base requirements • Journal is organized • Brings journal to class regularly	• Submits completed journal with base requirements and at least 1 additional article and 1 additional reflection (prompts to be provided) • Journal is organized and stylish (graphics, color, etc....) • Always brings journal to class and uses it as a resource
Work Ethic and Collaboration (10%)	• Misses two or more classes during unit (unexcused) • Regularly performs outside on contractual norms • Assignments completed beyond due dates • Is actively engaged in some phases of the project process • Sometimes participates in group activities	• Misses one class during unit (unexcused) • Rarely performs outside of contractual norms • Assignments completed by due dates • Is actively engaged in all phases of the project process • Always participates in group activities	• Attends all classes during unit • Always follows contractual norms • Assignments completed by due dates and provides aid to other students • Is actively engaged in all phases of the project process and is able to lead discussion when necessary • Is a facilitator of group activities
Individual Presentation (15%)	• Does not attend informal presentation with instructor • Attends solar cooking event dressed casually • Is disengaged from guests at solar cooking event • Presents journal to event guest and is rated as semi-prepared	• Attends informal presentation with instructor somewhat prepared to discuss content and project • Attends solar cooking event dressed professionally • Engages guests at solar cooking event • Presents journal to event guest and is rated as prepared	• Attends informal presentation with instructor and is prepared to discuss content and project • Attends solar cooking event dressed professionally and smiling • Invites guests to event and mingles during event • Presents journal to event guest and is rated as very prepared
Cooker Construction and Functionality (5%)	• Cooker cannot heat soup to 165°F within two hours • Cooker is unstable and tattered • Cooker cannot be constructed within budget	• Cooker heats soup to 165°F within two hours • Cooker is stable • Cooker is constructed with given budget	• Cooker heats soup to 165°F within one and a half hours • Cooker is stable and attractive (clean lines, attention to detail, decorations, etc...) • Cooker is constructed below budget
GROUP GRADE			

Sample Template 4.3: Solar Cooking with Conics Project Planning Form

Name of Project: *Solar Cooking with Conics*

Designed by Teacher Name(s) and Email Address(es): *Jacob Goodman, jag630@gmail.com*

Project Idea
What is the issue, problem, or theme of this project?

Bringing conic sections to life while learning about sustainability.

Topic(s) Addressed
List one or more topics this project addresses.

Parabolas, paraboloids, conic sections, importance of a focus and directrix.

Essential Question
What is the Driving Question or challenge?

As sustainability scientists, how do we use the properties of conic sections to serve hot soup outside on a winter day with no electricity or heat source other than the sun?

Entry Event
What is the hook to launch this project?

The students read a letter from Nate Roberts, the community partner from Pogue's Run Cooperative Grocer. We then read the letter as a class and discuss key components. After some discussion about the rubric, we watch a National Geographic video about solar cooking and reflect on it in writing.

CCSSM and SMP
List those to be addressed by the project

CCSSM
- HSA.SSE.B.3.A: Factor a quadratic expression to reveal the zeros of the function it defines.
- HSA.SSE.B.3.B: Complete the square in a quadratic expression to reveal the maximum or minimum value of the function it defines.
- HSF.IF.B.4: For a function that models a relationship between two quantities, interpret key features of graphs and tables in terms of the quantities, and sketch graphs showing key features given a verbal description of the relationship.
- HSF.IF.C.7: Graph functions expressed symbolically and show key features of the graph, by hand in simple cases and using technology for more complicated cases.
- HSF.IF.C.8.A: Use the process of factoring and completing the square in a quadratic function to show zeros, extreme values, and symmetry of the graph, and interpret these in terms of a context.
- HSG.GPE.A.1: Derive the equation of a circle of given center and radius using the Pythagorean theorem; complete the square to find the center and radius of a circle given by an equation.
- HSG.GPE.A.2: Derive the equation of a parabola given a focus and directrix.
- HSG.GPEA.3: Derive the equations of ellipses and hyperbolas given the foci, using the fact that the sum or difference of the distances from the foci is constant.
- HSG.GMD.B.4: Identify the shapes of two-dimensional cross-sections of three- dimensional objects, and identify three-dimensional objects generated by rotations of two-dimensional objects.
- HSG.MG.A.1: Use geometric shapes, their measures, and their properties to describe objects (e.g., modeling a tree trunk or a human torso as a cylinder).
- HSG.MG.A.3: Apply geometric methods to solve design problems (e.g., designing an object or structure to satisfy physical constraints or minimize cost; working with typographic grid systems based on ratios).

SMP
- MP1: Make sense of problems and persevere in solving them.
- MP2: Reason abstractly and quantitatively.
- MP5: Use appropriate tools strategically.
- MP8: L

Sample Template 4.3: Continued

T = Taught; P = Practiced; A = Assessed	T	P	A	T = Taught; P = Practiced; A = Assessed	T	P	A
Learner Outcomes Note the 21st-century skills taught, practiced, and/or assessed in this project.							
Written communication	■	■	■	Technology literacy	□	□	□
Oral communication	■	■	■	Work ethic	■	■	■
Collaboration	■	■	■	Civic responsibility	□	□	□
Critical thinking	■	■	■	Numeracy	□	■	□
Information lLiteracy	□	□	□	Core content skills	■	■	■

Habits of Mind Indicate one or two habits of mind that are the focus of this project.	□ Persisting □ Managing impulsivity □ Listening to others □ Thinking flexibly	□ Thinking about thinking ■ Striving for accuracy and precision □ Questioning, posing problems □ Applying past knowledge	□ Communicating with clarity □ Gathering data, using all senses ■ Creating, imagining, innovating ■ Responding with awe	□ Taking responsible risks □ Finding humor □ Thinking interdependently □ Learning continuously

Presentation Audience

Student Production		
Group products (major types):	• During the beginning of the project, students organize the entry documents and post NTKs through a Web 2.0 tool. • During the middle of the project, students work in data-driven groupings to remediate skills cooperatively. They also engage with GeoGebra software to begin to solve complex problems in the Challenge Packet. Ideally, they find a way to lead, either as an elected manager or during group activities. They understand where the project is headed and what is expected of them. • By the end of the project, students plan and construct a working solar cooker and plan the solar cooking event. The solar cooker must heat soup to 165°F within two hours.	**Check all that apply:** □ Class ■ School ■ Community □ Experts □ Web (public) ■ Parents □ Other:
Individual products (major types):	• During the beginning of the project, each student reflects on the National Geographic video. • During the middle of the project, each student completes a take-home quiz and journal entries with two reflections. • Each student submits a project journal demonstrating exploration and mastery of conic sections. They demonstrate understanding of sustainability issues through written reflection. • By the end of the project, students prove mastery of conic sections on the final quiz. Their journal is organized, complete, and submitted on time with evidence of active participation throughout the project.	

Sample Template 4.3: *Continued*

Assessments and Reflection	**Rubric(s)** Check and describe all that apply for this project	☐ Multimedia presentation rubric	■ Other: *Preliminary plans/ outlines/ prototypes*
		■ Oral presentation rubric	■ Other: *Working solar cooker*
		☐ CCSS ELA and literacy writing rubrics	☐ Other:
		☐ School writing rubric	☐ Other:
		☐ School learner outcomes rubric	☐ Other:
	Assessment Type(s) Check and describe all that apply for this project	■ Quiz: *A rigorous take-home quiz.*	☐ Performance Assessment:
		■ Test: *In-class formative quiz/test near the end of the unit.*	☐ Notes review:
		☐ Essay:	■ Checklist: *A checklist is provided multiple times throughout the project to reflect the growing list of resources and activities.*
		☐ Online assessment:	☐ Concept map:
	Reflection tools Check and describe all that apply for this project	■ Survey: *Students give honest feedback about project.*	☐ Focus group
		■ Discussion: *Debriefing is held to celebrate and to discuss possible mistakes.*	■ Personal learning plan: *Self-evaluation*
		■ Journal: *Students reflect after most lessons and keep all of their materials in one folder.*	☐ Student-teacher Conference
Project Resources	On-site personnel:	*Principal sign-off, aides for computer lab days (technology managers), computer lab*	
	Technology:	*One computer per student, basic tools (will depend on students' design)*	
	Community resources:	*Soup and supplies donation, sustainability expert (both provided by community partner), families, and community members as guests at event*	
	Print resources:	*Construction materials (will depend on students' design); string, straight edges, thumbtacks, etc, for drawing/exploring conics.*	
	Online resources:		

Template adapted from the Buck Institute for Education (www.bie.org).

Sample Template 4.4: *Solar Cooking with Conics Scaffolding NTKs*

	Scaffolding NTKs: Activity and Assessment Planning		
Anticipated Knowledge and Skills Students Need (NTKs)	**Assignment or Activity to Address NTKs**	**How Assignment or Activity Will Be Assessed**	**Learning Outcomes Addressed in Assignment/Activity**
What is a conic section?	Students are given a link to an interactive online applet that allows them to move a plane around two inverted cones. They explore the curves created by the cross-sections of the plane and cones and come to class with sketches of every shape they can create.	In groups of four, students will share the shapes they were able to create. Each group will compile their shapes and submit them. Groups will then share their shapes with the whole class.	This activity addresses the individual outcome of content mastery. It is students' first exposure to conics.
What is a solar cooker?	In the first lesson, students watch a National Geographic video clip on solar cooking and write reflections about the video.	Each student will source one article or webpage on solar cookers. They will add it to their journal and then use it later when planning and constructing the cooker.	This activity addresses the group outcome of constructing a working solar cooker
What is sustainability?	During our fifth lesson, students read a UN report on global sustainability. Our community partner will also speak to the students about sustainability.	This will be assessed through written reflection, which students will then include in their journals.	This activity addresses the individual outcome of demonstrating understanding of the concept of sustainability.
What is a cooperative grocery?	Our community partner will speak directly to the students about Pogue's Run Cooperative Grocer, sustainability, local food, etc.	This will be assessed by having students write a thank-you letter to our community partner that includes key concepts presented by that partner. Students then vote on the best letter, which will be sent to the partner.	This activity addresses the individual outcome of demonstrating understanding of the concept of sustainability.
How will we get the materials for our solar cooker?	This will be addressed directly during the NTK session. Students will be told that I will obtain the materials and that they will have a budget. Students will be given that budget and a list of places where the materials could be obtained (e.g., Lowe's, a thrift store) when they begin their preliminary planning.	This will be assessed during the planning phase of the construction of the solar cooker. As a class, students will submit a list of materials needed with expected costs. They will receive feedback from me regarding feasibility.	This activity addresses the group outcome of constructing a working solar cooker

Template adapted from the Buck Institute for Education (www.bie.org).

Sample Template 4.5: *Solar Cooking with Conics Six Problem-Solving Phases Planning Worksheet*

PBL Process Plan	NTKs	Scaffolding	Before Moving to Next to next Phase
Phase 1 What is the need/ problem?	As noted in Scaffolding NTKs, students will need to know what a conic section is, what sustainability means, what solar cookers are, and what their budget is. .	• Intro to conics workshop • Manager nominations	Students can complete the first written reflection.
Phase 2 What needs to be in our solution?	As this may be the most developed rubric students have ever worked with, they will need a mini workshop (at least) to analyze, discuss, and question the rubric. They will need clear expectations and deadlines.	Rubric workshop	At the end of the workshop, students will be divided into three sections and be asked to describe an A, B, or C student's work.
Phase 3 What are possible solutions?	A parabolic cooker is the only solar cooker (that I know of) created with conic sections. However, there are many types of parabolic cookers. Once students have reached the conclusion that a parabolic cooker should be used, they will be asked to research and bring in plans for a parabolic design	Google Docs and Apps workshop	In groups, students will compare their researched plans and choose a best plan based on budget, ease of construction, etc.
Phase 4 What solution should we use?	Each group will informally present their best choice for a parabolic solar cooker design. As a class, they will choose one based on budget, ease of construction, etc.	Coaching with design crew on budget and practicality.	The class will submit their plan with expected cost, tools, time needed, etc.
Phase 5 How do we create, run, and inspect our solution?	Students will need feedback from me regarding the feasibility of their design. Once their design is approved, students will need me to collect and pay for the materials. They will need time to actually construct the cooker. Finally, they will need a sunny day and an audience..	Final vote through Google Docs	Students will be present during the solar cooking event, journals in hand, and they will converse with the audience and informally present their exploration, process, and solution.
Phase 6 Reflect back: Did we solve the problem? Do we have a new problem?	Students will need regular feedback from me. I will assess their written reflections and quizzes. They will need structured group activities and whole-class discussion.	Each students gets a marked rubric with comments.	1. There will be a final all-class discussion about the project. 2. Students will have a final, longer, written reflection (with a rubric).

Template adapted from the Buck Institute for Education (www.bie.org).

Sample Template 4.6: Supplement to *Solar Cooking with Conics* Project Calendar

Solar Cooking with Conics

Overview: This PBL unit addresses the Common Core State Standards on conic sections while engaging students in the process of building a functional parabolic solar cooker and learning about local and global sustainability. Depending on class size, access to materials, and time available to commit to the construction of the cooker, group sizes could range from four students to whole-class groupings. The beginning, middle, and end of the unit have ordered sequences, while the lessons in between those parts are somewhat flexible and can be co-planned with students, families, and colleagues. Students should have regular access to technology in and out of class.

Pre-launch:

- Find one or more community partners—ideally, people who can aid in the PBL process in some way. Suggestions: an expert in sustainability, someone who can donate soup and other supplies, someone who can donate solar cooker materials, or a parent or community expert with knowledge of solar cooking.

- Create a shared Web-based document station with collaborative abilities, and keep it updated.

  Note: Google Drive meets all the needs of this project as outlined in this facilitation plan and in these documents.

- Create a physical document station that includes a place to organize documents. Post the Driving Question, and provide an NTK bucket (see fig. 4.5). Have copies of all documents available.

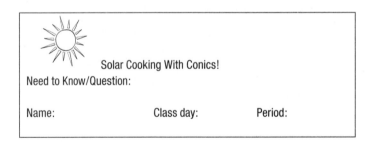

Fig. 4.5: Sample NTK slip.

Sample Template 4.6: *Continued*

Make it fun! Create an engaging space in the classroom dedicated to the project.

Note: I used a filing folder that contained a bank of resources for students if they needed additional help or wanted to access the previous days' materials, which proved to be very helpful.

- Secure project journals for students who may be unable to acquire them.
- Communicate with colleagues, families and students about project. Invite collaboration.

Launch:

1. Entry Event: Distribute and read the letter from community partner outlining the project.
2. Give a workshop on rubric, procedures, and expectations. Develop a contract. Elect a project manager. Have students form small groups and choose roles.

A sample student contract can be found on the NCTM website (www.nctm.org/more4u).

3. Distribute materials: letter to family (outlining the project and inviting them to the final event), objectives log (see fig. 4.6), journals.

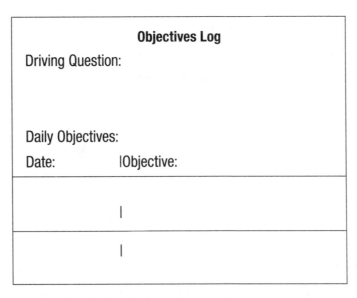

Fig. 4.6: Example of an objectives log.

Sample Template 4.6: *Continued*

4. Show National Geographic video; have students complete written reflection.
5. Periodically conduct journal checks, which helps to ensure that students stay organized.

> Remember: One of your three (or more) reflections must be from the Challenge Packet Prompts listed in the Reflection Prompts document.
>
> **We will have journal checks on Monday and Tuesday.** As always, you need your book and journal. 10 points for an up-to-date, organized journal with all documents including at least one reflection, two definitions illustrations, and Challenge Packet #1 and #3 completed.

Fig. 4.7: Example of teacher's announcements to the class.

6. Give students a chance to become familiar with Conic Section Explorer (an NCTM Illuminations applet) and other physical models of conics, and conduct a "What Are Conics?" workshop.
7. Distribute materials: blank notes, Challenge Packet.
8. Explore definitions (e.g., circle, ellipse, hyperbola, parabola) with GeoGebra, and conduct the drawing activity.
9. Scaffold Challenge Packet problem 1.
10. Distribute take-home quiz. Explain project submission procedures and answer questions.

Between the launch and middle of the project (these can go in almost any order):
- Conduct journal checks
- Scaffold the Challenge Packet problems with GeoGebra
- Learn the math standards
- Conduct workshop with community partner
- Collaborate with student groups, other classes, community members, etc.
- Reassess NTKs
- Provide a take-home quiz

Middle:
1. Ensure that all students have access to a computer or other device to submit take-home quiz answers.
2. Grade quizzes and process data. Create groupings based on data, and structure collaborative, student-driven reteaching of content as necessary.
3. Collect solar cooker designs. Facilitate voting on design through Web 2.0 collaboration.

Sample Template 4.6: *Continued*

> Send solar cooker designs to me by the end of the day Wednesday. **Your budget is $20.** I already have a lot of cardboard and reflective tape. You have $20 for any other needed materials. I will upload submitted designs Wednesday night. Electronic voting in Google Docs will be open all day Thursday and Friday and will count as a homework assignment. **Send designs to my email address.**

Fig. 4.8: Example of teacher's announcements to the class.

4. Facilitate sign-ups for building sessions.

Between the middle and end of the project (any order):

Relationship: Allow students to dictate some of the day-by-day activities once they understand where they are headed.

- Scaffold building process through in-class discussion and comparison of teacher's model with design plans
- Conduct journal checks
- Conduct peer assessment of journals
- Do I like/I wonder activity
- Review project rubric and expectations

> **Note:** Do this at just past the halfway point of the project, when students will have a better sense of their questions and concerns about the project.

- Provide organization time
- Scaffold Challenge Packet problems
- Find focal length of solar cooker
- Reteach concepts most commonly missed on the take-home quiz

Sample Template 4.6: *Continued*

End:

1. Give students the final journal checklist.

What Should Be In Your Journal

✓ Letter from Nate Roberts

✓ Project rubric

✓ Driving Question and daily targets

✓ National Geographic reflection

✓ Conic sections notes (filled out)

✓ Challenge Packet, **COMPLETED**

✓ Your class's manager list and Need-To-Knows

✓ Definitions illustrations

✓ **MWF Classes:** Challenge Packet #3 Guided Questions

✓ Indiana Standards sheet (filled out)

✓ Quiz 1 questions

✓ Quiz 1 corrections

✓ 2nd reflection, your choice

✓ Notes on finding the focal length of our solar cooker

✓ 3rd reflection, your choice

✓ UN article

✓ Solar cooker design plans

✓ Additional reflections, articles, and research

✓ "I like…" and "I wonder…" Post-It notes from a friend

✓ **TTh Classes:** April 16th warm-up

Fig. 4.9: Sample final journal checklist.

2. Conduct informal presentations with facilitator.

3. Have students sign up for a role in the event.

4. Administer final quiz.

5. Conduct solar cooking event.

6. Have students submit their journals.

Sample Template 4.7: Solar Cooking with Conics Project Calendar

Monday	Tuesday	Wednesday	Thursday	Friday
		WEEK 1		
• Students will understand project demands and develop contract • NTK workshop • National Geographic video and reflection		• Students will be able to identify each conic section • Students reflect on local sustainability and cooperative businesses • Community partner and "What Are Conics?" workshops • Students receive Challenge Packet		• Students will be able to interpret conics definitions with the help of GeoGebra • Take-home quiz • Journal check
		WEEK 2		
• Students will be able to develop equations of circles and parabolas • Take-home quiz scaffolding • Challenge Packet scaffolding		• Students will be able to propose solutions to Driving Question • Take-home quiz scaffolding • Challenge Packet scaffolding with GeoGebra		• Take-home quiz due • Students narrow down their solutions to one choice and develop plans for construction • United Nations article workshop
		WEEK 3		
• Students will be able to finish half the construction of their solar cooker • Students informally present project journals to instructor • Collaborative groupings based on take-home quiz data		Extra work day, if needed. • More scaffolding, reteaching, project work time, etc. • Time for students to organize journals and complete missing assignments		• Final quiz • Project reflection • Journals due at solar cooking event Note: This event is held sometime the following week, outside of class time

CHAPTER **5**

Interest in Interest

Crystal Collier, *Ben Davis University High School, Indianapolis, Indiana*

Buying a first car is a rite of passage for most young people in the United States. The freedom associated with finally being able to go where we want, when we want, as well as the social implications of having our own vehicle—these are things that most high school students dream about. This means that a project on car buying has instant relevance for students.

The inspiration for this unit came in large part from the teacher's personal experiences working in the financial industry. Over the years, she got to see firsthand just how little financial education most people actually have. Many of her clients had no understanding of basic financial principles, such as simple and compound interest; they had limited or no financial vocabulary; and they were generally unable to use data to make better financial choices—instead making choices based solely on emotional impulse.

This PBL unit was designed for an Algebra 2 course at Ben Davis University High School (BDU), an early college high school focusing on at-risk students, particularly those who are the first in their families to attend college. In the BDU environment, the unit took thirteen one-hour class periods to complete, including a summative assessment. The unit capitalizes on students' pre-existing enthusiasm for car ownership, and uses the idea of shopping for a car as a framework for learning about exponential growth, decay, and logarithms.

Though it was developed for an Algebra 2 class, the financial literacy in this unit is relevant to many age groups; this unit could also readily be adapted for other high school and college mathematics courses where exponents and logarithms are taught.

The overarching goal of *Interest in Interest* is that by the end of the project students will not only understand exponents and logarithms, but they will also understand how to use these concepts as part of a larger approach to determining a car's overall affordability. The unit leads students through planning and budgeting to buy a car that fits their needs (as differentiated from their *wants*) in terms of lifestyle and affordability. Working in groups of three or four, students apply exponents in analyzing the impact of interest on savings, loans, and depreciation.

(handwritten margin note: Choosing student groups)

Relationship: While there are many ways to form student groups, this teacher's favorite is student choice with a rationale: Prior to the Entry Event, students list their preferred group members, with a sentence or two explaining why they would like to work with each student. Remind the students to think carefully about who they indicate on their lists and to really consider what the other students will bring to the table. While the teacher still has final say regarding the groupings, this gives students some ownership of their group.

The information students need for this project is given through a series of in-class workshops combined with lab time for the students to do research and to develop spreadsheet models for their analyses. There is formative assessment of students' mastery of the mathematics, based on the worksheets they turn in (see fig. 5.1) and their practice presentation.

1. Determine whether this function is one-to-one:

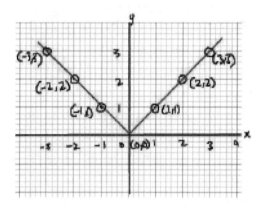

    Scale: On *x*-axis, 1 cm = 1 unit; on *y*-axis, 1 cm = 1 unit

2. If the following defines a one-to-one function, find its inverse. If not, write "Not one-to-one."

    $f = \{(2, 10), (2.5, -2.5), (6, 6), (8, 6)\}$

3. A man invests $3,000 in an account that pays 4% interest per year, compounded continuously.

    a. What amount of money will he have after 10 years?

    b. What kind of function is used to model interest growing in a bank account?

Figure 5.1. Sample student worksheet.

(handwritten margin note: graphs (linear vs. non-linear) for simple vs. compound vs. compound cont.)

At the conclusion of the project, each group makes a professional presentation of its research, using graphs and supporting calculations to explain the choice of a particular automobile. Combined with a traditional test on exponents and logarithms, this becomes a summative assessment.

The part of the project with which the BDU Algebra 2 students struggled the most was the presentation. Adding an extra peer evaluation day where the students presented to another group and received feedback from these "criticial friends" (see fig. 5.2) helped them refine their presentations and gave them more confidence for their final presentations.

peer review/ feedback

Critical friend: A person who can be trusted to offer honest and candid feedback in a constructive manner.

Circle YES (2 points) or NO (1 point).   Group #: _____

1. The presentation included at least two accurate graphs showing car depreciation.

                YES   NO   ___ points

2. The group was able to answer the questions asked by the panel.

                YES   NO   ___ points

3. The group made use of data to support the choice of a vehicle.

                YES   NO   ___ points

4. The group picked a vehicle that made sense based on the data.

                YES   NO   ___ points

5. The group maintained a professional appearance during the presentation.

                YES   NO   ___ points

                Total points: _____

Figure 5.2. Sample peer presentation rubric.

Students presented in front of a panel of judges, comprising school administrators, math instructors, and financial experts from the community. The average score on the presentations given by the panel was 90 percent proficiency. The audience also made many positive comments. For example, one math teacher stated,

> [This project] was extremely engaging for our high school students, ready to step into their next phase of life. They were challenged with the mathematical content of compound interest, displaying their collected data graphically, and compared various scenarios. The students found great value in the project, as they realized they would be in the market for a car purchase very soon if not already. The presentation was a great experience for the students, as they had to provide the rationale and support for their decision. Overall the PBL unit was practical, engaging, and supplied a healthy dose of meaningful content for our BDU early college high school students.

To conclude the project, students shared their successes and suggestions through a combination of written reflections (see fig. 5.3) and shared group reflections.

Relationship: At the end of the project, give students some time to reflect on their learning. How well did they work together? If students were allowed to choose their groups, were they satisfied with the groups? What would they change?

One thing I really liked about this project was:

One thing that I think would have made the project better is:

We used the following math concepts in this project:

I **DID   DID NOT**   find the math used in this project relevant to my life, because:

I learned the following mathematical and financial vocabulary words during this project:

I **DO   DO NOT** feel like I would be more comfortable using financial terms in a real-life setting as a result of this project, because:

I **DO   DO NOT**   think that this project has influenced how I might make a choice regarding a car in the future because:

Additional comments:

Figure 5.3. Sample reflection prompts.

[handwritten margin note: teach students how to research / find reliable online info]

In terms of academic success, the students who completed this project performed at the same proficiency level as those students who had been taught the material in a more traditional format. The main difference was that the students who completed the project seemed to have a better sense of how to apply exponents and logarithms than their traditional-instruction counterparts. In addition, many PBL students said that they had not previously known that there were reliable sources on the Internet for car-pricing data, and they would now use these sites to check vehicle prices if they, or someone they knew, were car searching.

Relevance: In fact, in the weeks following the project, several students noted that they had been able to help friends or family members with car shopping by sharing the resources and math techniques they had used. For an educator, this was truly the most satisfying aspect of the project!

Standards Addressed

This unit focuses on the use of exponents and logarithms, particularly on the concept of exponential growth and decay. Students also complete a great deal of work involving the creation and interpretation of their own functions, and they interpret functions that arise in applications in terms of a context.

Many of the CCSSM are addressed in this unit. Students have to select appropriate models (HSF. IF.B.4–6), analyze functions using different representations, focus on using key features to guide the selection of an appropriate type of model function (HSF.IF.C.7.B, HSF.IF.C.7.C, HSF.IF.C.7.E, HSF.IF.C.8, and HSF.IF.C.9), and construct and compare linear, quadratic, and exponential models and solve problems (HSF.LE.A.4).

Multiple SMPs are addressed throughout the unit. Primarily, students are challenged to construct viable arguments and critique the reasoning of others (MP3). The entire project is centered on using data to make a sound financial decision. Students use appropriate tools strategically (MP5) by generating their own data through research and analysis, which they then use to determine the best car for them. While students should understand that "best" is subjective, they find that data can be used concretely to substantiate their choice. Students also model with mathematics (MP4) as they learn how to use their exponential decay model to determine how well a car would maintain its value over time. Additionally, students make sense of problems and persevere in solving them (MP1) and reason abstractly and quantitatively (MP2) by considering both their data-based analysis and their subjective preference when selecting a vehicle. Students have to determine what made a car desirable based on specific criteria and then justify that choice. They have to consider a number of cars before they find one that meets their standards and stands up to analytical scrutiny.

Project Highlights

The following lessons highlight how the *Interest in Interest* unit is rigorous, is relevant to students, and helps students build relationships with one another.

Detailed lesson plans and other resources can be found on the NCTM website (www. nctm.org/more4u).

Day 1: Entry Event

This project begins with a letter from a local bank (see Sample Template 5.1) asking students to help develop new lending tools and products for recent college graduates. The students, already sorted into their groups, read the letter and then begin to develop a list of "Knows" and "NTKs."

entry activity

Relevance: Let students know that their "Knows" and "NTKs" lists are living documents that will change as the project progresses. NTKs will become "Knows," and there will be new NTKs to address.

These lists should cover the entire scope of the project, including math content, financial or car-related info, logistics, and time management. Groups then share items from their lists with one another and generate a class list.

A few NTKs will be immediately addressed with the rubric (see Sample Template 5.2) and the student group contracts. These documents answer many of the students' practical questions, open up discussion, and lead to the creation of additional NTKs. Have students decide on their group roles and then create a list of group norms, expectations, and consequences. By letting students have ownership of the norms and expectations of their groups, they are more likely to hold one another accountable and to keep one another on task.

Group management [handwritten note in left margin]

Relationship: Once students have selected their roles, meet with them in their role groups to be sure they understand their tasks and responsibilities. Point out that the other students in this role can be resources as they work on their project.

Day 4: Affordability Workshop

Students research how much money a college graduate with a degree leading to their planned career makes at graduation. They estimate the amount of money available to them to spend on a car per month, based on the following questions:

1. What is your career?
2. What is the average starting yearly salary (i.e., the total salary before taxes are deducted) for this career?
3. What is your estimated yearly take-home pay? (Taxes and benefits are approximately 25%. Multiply your starting salary by 0.75 to get your take-home pay.)
4. What is your estimated monthly take-home pay? (Divide your yearly take-home pay by 12.)
5. A conservative budget for your car is 20% of your take-home pay. What is your monthly car budget?

Students must understand that this is only an estimated budget and that the payments on a vehicle may not be this exact amount. They also need to understand that the budgeted amount includes insurance, fuel, maintenance, vehicle registration fees, and other costs.

Once students have determined their monthly car budgets, they discuss as a class how affordability affects their choice of car.

Students then compare auto loan rates from at least three financial institutions, noting that rates vary by model year of car, the term on the loan, whether the seller is a dealer or an individual, and whether the interest rate is fixed or variable. Students look up the rates online, or they can call or email a financial institution.

Based on these rates, students use an online auto loan calculator to find the monthly payments for three cars for which they have filled out individual automobile information forms.

A number of auto loan calculators can be found online (see, for example, autoloancalculator.com, nankrate.com, edmunds.com, or the websites of financial institutions).

Students compare these payments with their monthly car budget and share their findings with the class.

Day 6: Depreciation and Research Workshops

Students compare the Kelley Blue Book values for a given car with a continuous compounding interest model for a given depreciation rate. After an introduction to depreciation, students look up the Kelley Blue Book values for the current year and the previous five years for the car they are interested in purchasing and then plot these values on a graph. On the same graph, they plot the values from the continuous compounding interest formula, $A = Pe^{rt}$, where A is the amount the car is worth after t years for some depreciation rate r and initial value P.

Typically, a car loses 15–20 percent of its value each year. Table 5.1 and figure 5.4 show how much a used car purchased for $10,000 is worth for five years of ownership, based on both the Kelley Blue Book values and the estimated value from the continuous compounding formula with a 15 percent depreciation rate.

Table 5.1. Depreciation of a car

Year	Kelley Blue Book	Estimate
0	$10,000	$10,000
1	$9,758.48	$8,607.08
2	$9,485.75	$7,408.18
3	$8,758.48	$6,376.28
4	$8,061.51	$5,488.12
5	$7,788.78	$4,723.67

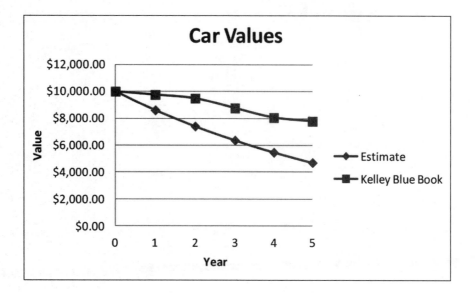

Figure 5.4. Car depreciation over time.

Day 12: Final Presentations

The unit concludes with students sharing what they have learned in a final presentation. One thing to note is that while table 5.1 shows that the Kelley Blue Book value outperforms the depreciation estimate, this is not the case for all cars; students' presentations may include a variety of scenarios. A final question to pose to students is, Could you sell your car after one, two, or three years and have enough money to pay back the loan?

Most BDU Algebra 2 students stated that this project influenced how they would choose a car in the future. One student commented, "Learning math like this was hard, but I think it was good. I think I understand it better because I can think about the project and remember how to do it." Another summed it up by saying, "I was way more interested in the math because I was buying my dream car."

This unit clearly demonstrates that integrating financial concepts into mathematics courses not only prepares students for other financial literacy courses but also adds relevance to the mathematics they are currently studying. Through the use of projects involving financial concepts and financial examples, this can be done with no loss of instructional time.

Sample Template 5.1: *Interest in Interest* Entry Event

To: BDU Math 102 Students
From: Preston Beale Jr., Assistant Vice President, P. Beale Savings & Loan
Re: Our New Student Ad Campaign

Dear Student:

As you know, modern consumers are subjected to a seemingly endless stream of advertisements and purchase pressures. So, it's more important than ever to learn how to live within your means, because your purchase choices now can greatly affect your financial future.

Buying a first car is a very exciting prospect to young people as they enter adulthood. Cars play a number of different roles in modern American culture. A car is a tool for both getting things done and for having fun, and is a symbol of self-expression. It's also probably one of the biggest purchases you've ever made, and the consequences of that purchase, good or bad, can affect you for years to come.

P. Beale Savings & Loan has a proud tradition of helping get responsible young adults like you behind the wheel with our competitive auto loans and supportive financial counseling. We are committed to helping you make a financial decision you can be proud of; but to do that effectively, we first need YOUR help. We need you to step into the role of a new college graduate looking to buy his or her first car and to answer the following question: How can we, as recent college graduates, determine the best vehicle purchase for our income?

By using a methodical approach and doing research, and using your own expected income level as a college graduate, come up with a budget, find a lender (it doesn't have to be us) with interest rates and monthly payments you can afford, and choose the best vehicle that fits that budget. By "best vehicle" we mean a car that is affordable, holds its value, and will serve both your needs and personal desires.

We then ask that you prepare a simple presentation showing both the vehicle you chose and the criteria that led to that decision. Use the data. We hope to help guide future college grads to making the best car-buying decision they can make. We also sincerely hope that by taking part in this project, you'll empower yourself to make your own great car purchase!

We look forward to your presentations and wish you good luck as you begin working. You have the opportunity to change lives through this project and we hope that you do.

Best wishes,

Preston Beale Jr.
Assistant Vice President
P. Beale Savings & Loan

Sample Template 5.2: *Interest in Interest* PBL Unit Project Rubric

Criteria	Junker (Below Standards)	Used Car (Acceptable)	Fresh off the Factory Line (Exemplary)
Content (10 Points):	• Students demonstrate little or no level of proficiency with mathematical vocabulary. • Students show some evidence of interpreting exponential and logarithmic functions that arise in applications. • Students analyze functions using at least one representation. • Included graphs are inaccurate and/or irrelevant to the project. • Students do not give accurate mathematical explanation of their graphs. • Students show some evidence of attempting to create a model to solve the problem of buying a car.	• Students demonstrate a basic level of proficiency with mathematical vocabulary. • Students interpret exponential and logarithmic functions that arise in the context of buying a car. • Students analyze functions using at least two representations. • Graphs are accurate and relevant to the proposal. • Students give accurate mathematical explanation of their graphs. • Students construct a mathematical model to solve the problem of buying a car.	In addition to Used Car criteria: • Students demonstrate a high level of proficiency with mathematical vocabulary. • Students interpret exponential and logarithmic functions that arise in applications in terms of the context. • Students analyze functions using multiple representations and use them to strengthen their proposal. • Students construct and compare linear, quadratic, and exponential models and solve problems.
	1 --- 5	6 --- 8	9 --- 10
Reasoning (10 Points):	• Students do not have a defined reason for their proposal. • Students present fewer than three pieces of data to support their proposal. • Data are not relevant to the proposal.	• Students have a well-defined rationale for their proposal. • Students present at least three pieces of data to support their proposal. • Data are relevant to the proposal.	In addition to the Used Car criteria, • Students use multiple pieces of relevant data to support their proposal.
	1 --- 5	6 --- 8	9 --- 10
Presentation (10 Points):	• Students do not make use of available technology. • Students incorrectly use mathematical or financial language to present their proposal. • Students are unable to answer questions regarding their proposal.	• Students use technology in their presentation. • Students correctly use mathematical and financial language to present their proposal. • Students answer questions regarding their proposal.	In addition to the Used Car criteria, • Students speak clearly and have good posture. • Students demonstrate professional behavior throughout the presentation.
	1 --- 5	6 --- 8	9 --- 10

Points: _____/30

Sample Template 5.3: *Interest in Interest* Project Planning Form

Name of Project: Interest In Interest

Designed by Teacher(s) and Email(s): *Crystal Collier, crystal.collier@wayne.k12.in.us*

Project Idea What is the issue, problem or theme of this project?	*Students will use a variety of mathematical data to support a potential car purchase*
Topic(s) Addressed List one or more topics this project addresses.	*Financial literacy*
Essential Question What is the Driving Question or challenge?	*As recent college graduates, how can we determine the best vehicle to purchase based on our income?*
Entry Event What is the hook to launch this project?	*The class will be broken into groups of three or four and given the entry document, which asks students to conduct a research project, in the guise of recent college graduates, to help a fictitious bank develop marketing materials that teach students about financial choices specifically related to purchasing a vehicle. Students will use both the entry document and the rubric to develop a list of "Knows" and "NTKs" for their group. This will be expanded to a classroom list that will be posted in the classroom for the duration of the project.*
CCSSM and SMPs List those to be addressed by the project.	CCSSM • HSF.IF.B.4: For a function that models a relationship between two quantities, interpret key features of graphs and tables in terms of the quantities, and sketch graphs showing key features given a verbal description of the relationship. • HSF.IF.B.5: Relate the domain of a function to its graph and, where applicable, to the quantitative relationship it describes. • HSF.IF.B.6: Calculate and interpret the average rate of change of a function (presented symbolically or as a table) over a specified interval. Estimate the rate of change from a graph. • HSF.IF.C.7B: Graph square root, cube root, and piecewise-defined functions, including step functions and absolute value functions. • HSF.IF.C.7C: Graph polynomial functions, identifying zeros when suitable factorizations are available, and showing end behavior. • HSF.IF.C.7E: Graph exponential and logarithmic functions, showing intercepts and end behavior, and trigonometric functions, showing period, midline, and amplitude. • HSF.IF.C.8: Write a function defined by an expression in different but equivalent forms to reveal and explain different properties of the function. • HSF.IF.C.9: Compare properties of two functions each represented in a different way (algebraically, graphically, numerically in tables, or by verbal descriptions). • HSF.LE.A.4: For exponential models, express as a logarithm the solution to $ab^{(ct)} = d$ where a, c, and d are numbers and the base b is 2, 10, or e; evaluate the logarithm using technology. SMP: • MP1: Make sense of problems and persevere in solving them. • MP2: Reason abstractly and quantitatively • MP3: Construct viable arguments and critique the reasoning of others. • MP4: Model with mathematics. • MP5: Use appropriate tools strategically.

Sample Template 5.3: Continued

Learner Outcomes
Note the 21st-century skills taught, practiced, and/or assessed in this project.

T = Taught; P = Practiced; A = Assessed	T	P	A	T = Taught; P = Practiced; A = Assessed	T	P	A
Written communication	■	■	■	Technology literacy	■	■	■
Oral communication	■	■	■	Work ethic	□	■	□
Collaboration	■	□	■	Civic responsibility	□	□	□
Critical thinking	□	■	□	Numeracy	■	□	□
Information literacy	■	□	□	Core content skills	■	■	■

Habits of Mind
Indicate one or two habits of mind that are the focus of this project.

□ Persisting	■ Thinking about thinking	■ Communicating with clarity	□ Taking responsible risks
□ Managing impulsivity	□ Striving for accuracy and precision	■ Gathering data, using all senses	□ Finding humor
□ Listening to others	□ Questioning, posing problems	□ Creating, imagining, innovating	□ Thinking interdependently
■ Thinking flexibly	□ Applying past knowledge	□ Responding with awe	□ Learning continuously

Presentation Audience

Check all that apply:
- ■ Class
- ■ School
- ■ Community
- ■ Experts
- □ Web (public)
- □ Parents
- □ Other:

Student Production

Group products (major types):

Beginning of project:
- Students are given the entry document and rubric. They use this to generate a list of "Knows" and "NTKs." This list will be revisited over the course of the unit.
- Once students are assigned to their groups, they read a group contract outlining the expectations for each group member. Students are given the opportunity to edit this document within reason if they choose. Once finalized, students sign the document and begin work on their projects.
- Groups begin by choosing a postgrad career and using the average income for their group's career as a basis for their financial situation. They select a vehicle to research and a bank through which to finance their purchase. Students use the Internet to conduct their research but are also encouraged to reach out directly to local business via email, phone, or visit.

End of project:
- Each group will make a professional-level presentation of research accumulated to support the purchase of a particular automobile including graphs and supporting calculations. They will be given a chance to rehearse their presentation in front of a peer group in order to receive feedback. Their presentations will be viewed by an authentic audience comprised of school administrators, math instructors, and financial experts.
- Students will be prepared to answer questions about the banking products featured in the research and be able to support their car choice with the research they conducted. (i.e. the value of the car remains stable over time, as good gas mileage, meets their needs, etc.)
- After the presentations students debrief the project for its content. They will also be given a chance to assess themselves.

Individual products (major types):

Middle of project:
- Students participate in a variety of workshops on car buying, exponents, financial products, technology, and presentation skills. They use the information from these workshops to further develop their case studies.
- Students are assessed on their learning of the content through a quiz and in their final presentations.
- Students also complete and turn in worksheets, which they can use in their final projects.
- Each student will be able to accurately solve a variety of problems involving exponential expressions. They will be able to speak knowledgeably when asked to explain the process they used to solve the various equations, and they will be able to accurately interpret the graphs of exponential equations.

Sample Template 5.3: Continued

Assessments and Reflection	Rubric(s) Check and describe all that will be used for this project.	■ Multimedia presentation rubric	□ Other:
		■ Oral presentation rubric	□ Other:
		□ CCSS ELA and literacy writing rubrics	□ Other:
		□ School writing rubric	□ Other:
		□ School learner outcomes rubric	□ Other:
	Assessment Type(s) Check and describe all that will be used for this project.	□ Quiz:	■ Performance assessment: Final projects are assessed by panel of experts including teachers and community partners
		□ Test:	□ Notes review:
		□ Essay:	□ Checklist:
		□ Online assessment:	□ Concept map:
	Reflection Tools Check and describe all that will be used for this project.	■ Survey: Student reflection questionnaire	□ Focus group:
		■ Discussion: Student debrief of the project outcomes.	□ Personal learning plan:
		□ Journal:	□ Student-teacher Conference:
Project Resources	On-site personnel:	Teacher, Media Center Specialist	
	Technology:	Internet access, computers with presentation software available (PowerPoint, Google Slides, etc.)	
	Community resources:	Banking professionals, car sales experts (sales persons, mechanics, etc.)	
	Print resources:	Newspaper advertisements, banking/lending brochures	
	Online resources:	KBB.com, Edmunds.com, cars.com, carfax.com	

Template adapted from the Buck Institute for Education (www.bie.org).

Sample Template 5.4: *Interest in Interest Scaffolding NTKs*

Scaffolding NTKs: Activity and Assessment Planning

Anticipated Knowledge and Skills Students Need (NTKs)	Assignment or Activity to Address NTKs	How Assignment or Activity Will Be Assessed	Learning Outcomes Addressed in Assignment/Activity
How will this be graded?	Rubric will be given and explained to the students.	Students will be given an opportunity to rewrite the rubric. Students will demonstrate understanding of the rubric by being able to transfer it into student-friendly language.	Students can explain, in their own words, the methods by which they will be assessed in this project.
What is a case study?	Students will be given a sample case study to review. We will discuss the structure of a case study and the methods used to generate and analyze relevant data.	Students will create an original case study using methods similar to those used in the example.	Students will create an original case study using methods similar to those used in the example.
What types of graphs do I need to include?	Students will participate in a workshop on graphing and developing meaningful graphs.	Students will be asked to use their case study to create at least three different graphs illustrating different savings strategies.	Students can create a graph to illustrate how the banking product works over time, given certain variables (e.g., the value of the car remains stable over time).
What is a professional presentation?	Students will participate in a workshop on developing a concept for their presentations and will receive feedback to help them develop that concept to a professional level.	Students will submit a rough draft or outline of their presentation with a prototype copy of the original marketing material to be presented. Students will have an opportunity to revamp their draft as needed based on any feedback they receive. Students will demonstrate a solid understanding of professional presentations through the execution of their own presentations and materials.	Students can present their ideas in a clear, concise, and professional manner
How do I use Excel?	Students will participate in a workshop on how to use Excel to generate data and create graphs for their project.	Students will complete an assignment focused on graphing and using Excel to create graphs. Their graphs must be accurate and relevant to the case study.	Students can use Excel to create a variety of graphs that illustrate various logarithmic and exponential functions.
How do I use PowerPoint?	Students will participate in a workshop on how to use PowerPoint to embed objects and create a presentation of their case study.	Students will effectively use PowerPoint to present their case study.	Students can create a PowerPoint slide show that engages the audience in their case study.
What math will I need to know?	Students will participate in two to three workshops on exponential functions and logarithmic functions, which will be addressed in the context of the project.	Students will take quizzes and a test to determine the level of learning they have achieved on these learning goals. Students will also demonstrate their learning through the final presentation by presenting accurate data that they are able to properly interpret. Students will be able to answer questions from the panel regarding their findings and will support their answers with mathematical data.	Students can accurately solve and interpret logarithmic and exponential equations.

Template adapted from the Buck Institute for Education (www.bie.org).

Sample Template 5.5: *Interest in Interest Six Problem-Solving Phases Planning Worksheet**

PBL Process Phase	NTKs	Scaffolding	Before Moving to Next Phase . . .
Phase 1 What is the need or problem?	• How do we create a presentation suitable for this project? • Which group members are responsible for specific tasks associated with the project?	• Group contracts • Identify which NTKs are necessary to identify the need or problem.	1. Students can explain the challenge or problem. 2. Students explain the problem to the facilitators.
Phase 2 What needs to be in our solution?	• What criteria should we use to support our vehicle choices? • What resources can we use to gather data on car values? • How do we create graphics such as graphs and tables for our presentations? • What other math is needed to complete the project?	• Student-created criteria checklist • Car buying Workshop • Affordability Workshop • Technology Workshop	1. Groups explain or defend their solution criteria (a student-generated checklist). 2. Groups compare and contrast their checklist with other groups' checklists.
Phase 3 What are possible solutions?	• What might a successful presentation look like? • How can we use graphs and tables in our presentation?	Critical friends feedback with peer teams	Students present multiple solution ideas based on criteria.
Phase 4 What solution should we use?	• How do we use data to support an overall choice? • How do we best present the data to support our case?	Rubric checks where students determine their score	Students articulate what solution they are going to do and defend it with the rubric.
Phase 5 How do we create, run, and inspect our solution?	• What issues did the critical friends see in Phase 3? • Should we address these concerns? How can we address them?	Critical friends feedback with peer teams	Students incorporate critical friends feedback into their revised PowerPoint presentation.
Phase 6 Reflect back: Did we solve the problem? Do we have a new problem?	• What does the group believe are the strengths and weaknesses of their presentation? • How did the panel feedback align with the group's self-assessment? • Does this car-buying simulation mirror real life? Why or why not?	Reflection survey and discussion	1. Expert/outsider checks student work against the rubric. 2. Students complete a reflection on what they learned.

Sample Template 5.6: *Interest in Interest Project Calendar*

Monday	Tuesday	Wednesday	Thursday	Friday
27	28	29 Entry Documents Knows/NTKs	1	2 Car Buying Workshop
5 Technology Workshop I (Car Buying Resources)	6	7 Affordability Workshop	8	9 Quiz Review Lab Time Post Lab Check-In
12 Quiz Technology Workshop (Using Excel and PowerPoint)	13	14 Depreciation Workshop	15	16 Lab Time Presentation Development
19 Checklist Check In	20	21 Presentation Practice Peer Review	22	23 Lab Time Project Checklist Review
26 Final Presentations	27	28 Project Debrief Content Review	29	30

Interest in Interest Sample Student Product

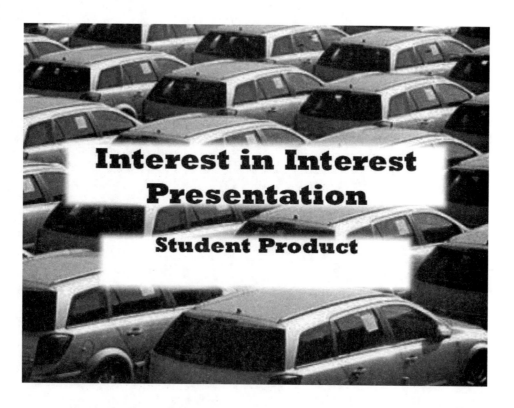

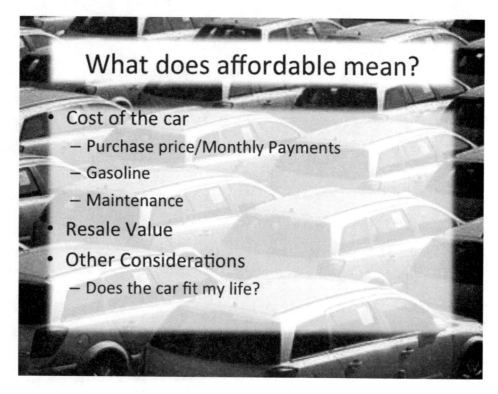

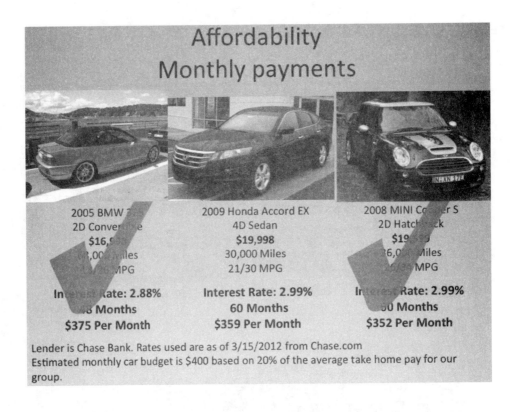

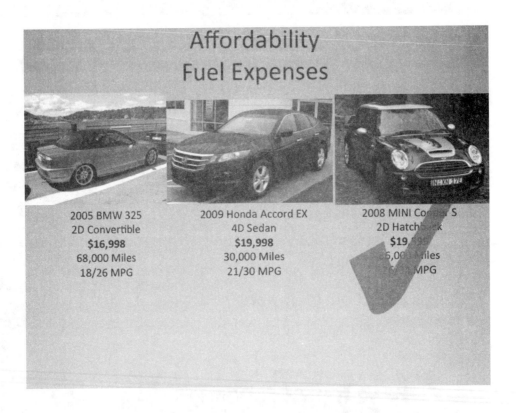

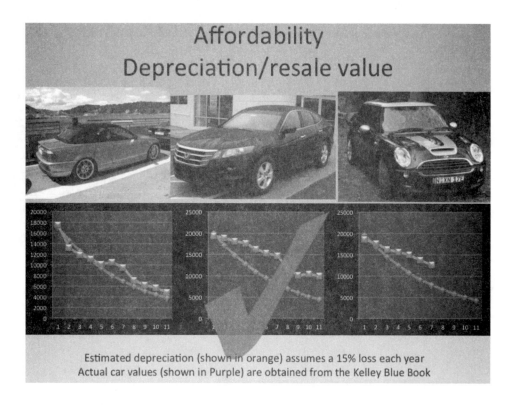

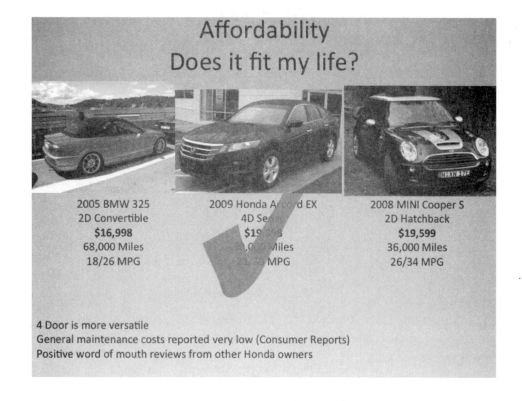

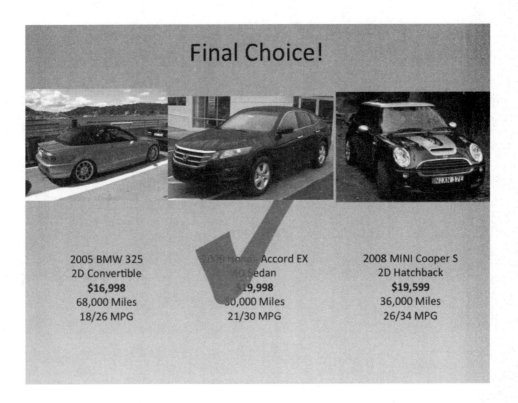

Cabriolet: By nakhon100 (BMW 325 Ci Cabriolet M Sport E46) [CC BY 2.0 (http://creativecommons.org/licenses/by/2.0)], via Wikimedia Commons

Honda Accord: By IFCAR (Own work) [Public domain], via Wikimedia Commons

Mini Cooper: By FearTec at English Wikipedia (transferred from en.wikipedia to Commons) [Public domain], via Wikimedia Commons

CHAPTER

6

Polyhedron Tiling Artwork Project

Maple So, *University of Indianapolis, Indiana*

This unit was created as the author was studying PBL in her graduate coursework. She really wanted to use art and geometry to capture the attention of her five geometry classes (a total of about 125 students), and she also wanted students to incorporate art into their geometric skills. During the early stages of designing the *Polyhedron Tiling Artwork Project*, she researched different package designs that are suitable for an egg-drop experiment, because scientific experiments usually involve repeated measurements. When she noticed the different geometric shapes that are involved in making a package design for an egg-drop test, her focus then shifted to studying just the geometric solids. She was also fascinated by Arabic tiling designs, which led her to look into various Archimedean tiling patterns to decorating the polyhedrons.

This PBL unit is different from others in that it does not involve a community partner. However, it is a great example of how to scaffold non-PBL learners into a PBL environment by implementing rigorous PBL practices in a mathematics classroom.

Standards Addressed

The *Polyhedron Tiling Artwork Project* engaged students' learning by focusing on three SMPs:

- Students **reasoned abstractly and quantitatively** (MP2) as they examined how three-dimensional (3-D) figures can be projected onto two-dimensional (2-D) space when students studied 2-D net drawings for each polyhedron.

- The unit pushed students to demonstrate their comprehension of volume and surface area as they built 3-D models of polyhedrons and tessellated each face of the solid with a pattern of polygons. They **modeled with mathematics** (MP4) by explaining how the surface area formulas correspond to the 2-D net drawings of various solids.

- Throughout the unit, students **looked for and made use of (geometric) structure** (MP7) as they tessellated each face of their polyhedron with polygon patterns that could not overlap or create gaps.

73

The project also addressed several Common Core content standards in geometry: Explain volume formulas and use them to solve problems (HSG.GMD.A.1, HSG.GMD.A.3), visualize relationships between 2-D and 3-D objects (HSG.GMD.B.4), apply geometric concepts in modeling situations (HSG.MG.A.1, HSG.MG.A.3), make geometric constructions (HSG.CO.D.13), and define trigonometric ratios and solve problems involving right triangles (HSG.SRT.C.8).

In addition, group members developed communication skills as they conveyed their ideas to one another. They learned how to allocate tasks among the group, based on interest and individual ability; how to organize tasks based on priority; and how to ensure that each group member did his or her job in the product creation.

Project Highlights

The Driving Question of this project is, "How can understanding the faces, vertices, and edges of convex polygons enable mathematicians to better comprehend the surface area and volume of geometric solids?" During the tessellation process, students can see the connections between math and art; they can see how repeated patterns in tessellation require spatial analysis in order to prevent gaps and overlaps that do not fit onto the faces of their solids. Students can also apply the concept of surface area to real-life situations as they apply concepts of faces, vertices, and edges of convex polygons. For example, students can explore the minimum amount of wrapping paper needed to wrap a 3-D figure. Other real-life applications include painting a house and putting up a tent.

Rigor: This unit is mathematically rigorous because it requires students to apply their knowledge of 2-D space to 3-D geometry. Students must understand the Pythagorean theorem and right-triangle trigonometry when they calculate the volume and surface area of their polyhedrons.

Day 1: Entry Event

The Entry Event had two components: Students constructed geometric solids from flattened cutouts and then analyzed the layouts of each solid:

- In groups of four, students assembled a triangular prism, a rectangular prism, a pentagonal prism, and a square pyramid.
- They analyzed each solid by describing the shape of each polygon that forms the faces of the solid and finding the total number of vertices, edges, and faces for each solid. (See fig. 6.1.)

tie-in to ND ch. 10

Getting Started: Exploring Polyhedrons
Complete the table for each solid below. Use the 2-D net cutouts to fill out the tables.

Draw the 2-D view of the polyhedron.		Draw the 3-D view of the polyhedron.	
What is the name of the polyhedron?		Answer:	
Name and draw all the polygons that form the faces of the solid.		Answer:	
Total vertices:		Total edges:	Total faces:

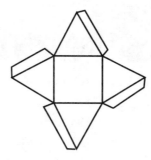

Fig. 6.1: Snapshot from Entry Event worksheet.

The Entry Event concluded with each student writing a hypothesis of what *surface area* and *volume* meant in their own words. (See Sample Template 6.10 for more detailed lesson plans for this project.)

Additional resources for this unit can be found on the NCTM website (www.nctm.org/more4u).

Skills Development

Students' spatial analysis skills grew as they struggled to make multiple drafts of their polyhedrons. The teacher enjoyed seeing the students formulate questions about how to find the area of irregular polygons, learn how to partition irregular polygons into smaller polygons during the calculation process, and research how to find the apothem of a regular pentagon. She was continually impressed by her students' creative approaches—for example, one student used a compass to construct congruent angles from a circular base so that she could properly cut out a regular pentagon using the length of the apothem.

Students had the opportunity to demonstrate a number of 21st-century skills as they collaborated with their team members in constructing their polyhedrons and presenting their final product. Sample Templates 6.7, 6.8, and 6.9 show how the teacher scaffolded students to collaborate and exercise 21st-century skills. During their collaboration, students used their Google accounts to share files and communicate outside of school.

Rigor: This integration of technology required students to learn skills in time management, communication, and division of labor among teammates inside and outside the classroom as they combined their coursework files in the shared Google files. These are all important 21st-century skills and are key to college readiness.

To conclude the unit, students gave a formal presentation of their final polyhedrons with tessellated faces (see fig. 6.2).

Fig. 6.2. A group's final polyhedron.

Their presentations included their calculations of surface area and volume, which required them to use mathematical vocabulary to show their mastery of the content (see fig. 6.3).

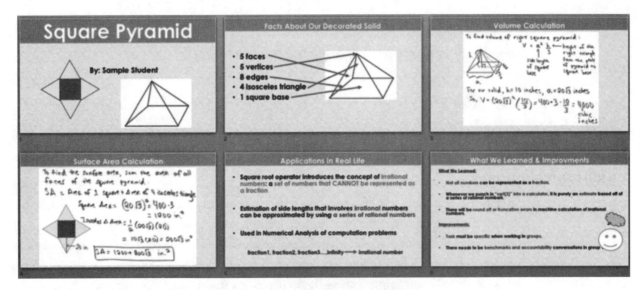

Fig. 6.3: Student slideshow presentation.

Ultimately, this project fostered students' readiness for college and careers in a number of ways:

- Students developed their written and oral communication skills throughout the unit, as they learned to communicate their needs and questions in a professional way to their teacher and peers.

- Students learned to manage their time and meet smaller project benchmarks before creating the final product. They practiced dividing a large project into smaller tasks that were doable in the time immediately available. As they completed these tasks, they brought questions to the teacher on how to create the tessellation patterns, calculate the volume and surface area, and integrate mathematical vocabulary in their presentations. The students asked necessary questions in a timely fashion as they met their project benchmarks; they did not wait till the last minute.

- Students practiced working as part of a team in building the tessellated polyhedron, planning, communicating, and problem-solving with their small-group members during their collaboration time.

Differentiated Instruction

During the unit, the teacher used differentiation to meet students' different levels of math ability. For students who struggled deeply in math, the teacher assigned a simpler solid, such as a cube or rectangular prism. For students with higher math ability, the teacher assigned polyhedrons that had more faces and edges, such as an octahedron.

Rigor: PBL is student-centered and allows teachers to meet the needs of students in a variety of ways. Think about how to differentiate the content, product, and process for students.

The teacher made 2-D nets of different solids to help students see the flattened view of 3-D solids. She also used Interactive Geometry (Annenberg Foundation 2016), an online application, to help students see 3-D objects from different perspectives. She created guided notes and pictures of polyhedrons, with their cross-sectional views already drawn, to help students break down volume and surface area problems.

Relevance: The teacher used everyday packaging items, such as cereal boxes, cheese cube boxes, and soup cans, to make learning more tangible and applicable to students' daily lives. Allowing students to take apart packaging materials and flatten containers really gave them a solid understanding of 3-D space.

Final Assessment

At the end of the unit, students completed peer evaluations and a self-evaluation for the project. They provided a rationale for the evaluation points they gave to themselves and to their teammates. Students also wrote short answers to questions about what they struggled with, what they enjoyed, and what improvements to the project they would recommend. The teacher compiled all written responses into a piece of word art, using Wordle, to show students what the class as a whole thought about the project.

Relationships: Students used Google Forms to "grade" themselves and their teammates online. Doing this privately meant that they were not pressured to give specific evaluation scores during class, which seemed to increase their comfort with assessing their peers' work.

Challenges and Suggestions for Future Iterations of the Project

Students' time management and self-regulation were a challenge for the teacher. Some groups completed the project benchmarks more swiftly than others, while others took a lot of time to make final decisions on making measurements for their polyhedrons.

The teacher was also challenged by allowing students to self-regulate their project progress. Several groups struggled with student leadership within their teams, such as deciding on the presentation details and documenting their daily tasks in a written log. The teacher had to learn to allow students to resolve project- management issues on their own and to interrupt only to offer advice, rather than give instructions.

One suggestion for teachers who want to launch a successful project is to provide agendas of all the weekly benchmarks and assessments ahead of time to the students. This will help them properly manage their time between creating the solid and studying all the geometric topics covered during a specific week.

Another tip is to provide students with an equation sheet for the different volume and surface area formulas (see fig. 6.4). This will remove the misconception that the formulas must be memorized and will help students focus on their actual task: applying the formulas to create a final product.

Make formula sheet less intimidating ↓

Surface Area Formulas

Cylinder: $2\pi r^2 + 2\pi rh$

Cone: $\pi r^2 + \pi r *$ (Slant height)

Regular pyramid: Area of base + 1/2 (Perimeter of base) * (Slant height)

Sphere: $4\pi r^2$

Volume Formulas

Prism: Area of base * (Height)

Cylinder: $\pi r^2 h$

Regular Pyramid: 1/3 * Area of base * (Height)

Cone: $1/3 * (\pi r^2 h)$

Sphere: $4/3 * \pi r^3$

Fig. 6.4: Sample formula equation sheet.

Concluding Thoughts

The teacher noted problems that students missed most often on the final assessments, and she subsequently edited her guided notes and assignments to help break down the problems into smaller questions. This will help content delivery for future classes because the teacher is now aware of the aspects of the mathematical content that students struggled with.

The teacher used Google Forms to have students evaluate the quality of the project and teacher instruction, which provided the teacher with instantaneous data on what students thought went well in the unit and the aspects of instruction that needed improvement. Through the feedback she received on this and other project implementations, the teacher thinks that her PBL units are getting better.

PBL allowed the teacher to understand that a good project requires ongoing feedback from the students and that constant revisions strengthen lesson planning in project units.

References

Annenberg Foundation. *Interactives: 3-D Geometry Shapes.* St. Louis, Mo.: Annenberg Learner (2016), http://www.learner.org/interactives/geometry/area_surface.html.

Sample Template 6.1: *Polyhedron Tiling Artwork Project* Planning Form

Name of Project: *Polyhedron Tiling Artwork Project*

Designed by (Teacher Name[s] and Email Address[es]): *Maple So, somaplemail@yahoo.com*

Project Idea What is the issue, problem, or theme of this project?	*The theme of this project is to build a polyhedron and tessellate its faces with patterns of polygons. Students find the volume and surface area of their solids. They use the surface area of their polyhedron to determine how to partition their polyhedron's faces with patterns of polygons. Each tessellated face may not have gaps or overlaps.*
Topic(s) Addressed List one or more topics this project addresses.	*This project addresses three-dimensional geometry, right-triangle trigonometry, volume, surface area, and algebra.*
Essential Question What is the Driving Question or challenge?	*How can understanding the faces, vertices, and edges of convex polygons enable mathematicians to better comprehend the surface area and volume of geometric solids?*
Entry Event What is the hook to launch this project?	*The Entry Event consists of two components: construction of geometric solids and analysis of the layouts of the different solids.* • *Students work in groups of four to assemble a triangular prism, rectangular prism, pentagonal prism, and square pyramid from the flattened cutouts they are given for each geometric solid.* • *Students describe the shape of each polygon that forms the faces of the solid and find the total number of vertices, edges, and faces for each solid. They then hypothesize what surface area and volume mean in their own words. Their analysis will be individually written and submitted.*

Sample Template 6.1: Continued

CCSSM and SMP	CCSSM:
List those to be addressed by the project.	• HSG.GMD.A.1: Give an informal argument for the formulas for the circumference of a circle, area of a circle, volume of a cylinder, pyramid, and cone.
	• HSG.GMD.A.3: Use volume formulas for cylinders, pyramids, cones, and spheres to solve problems.
	• HSG.GMD.B.4: Identify the shapes of two-dimensional cross sections of three-dimensional objects, and identify three-dimensional objects generated by rotations of two-dimensional objects.
	• HSG.MG.A.1: Use geometric shapes, their measures, and their properties to describe objects (e.g., modeling a tree trunk or a human torso as a cylinder).
	• HSG.MG.A.3: Apply geometric methods to solve design problems (e.g., designing an object or structure to satisfy physical constraints or minimize cost; working with typographic grid systems based on ratios).
	• HSG.CO.D.13: Construct an equilateral triangle, a square, and a regular hexagon inscribed in a circle.
	• HSG.SRT.C.8: Use trigonometric ratios and the Pythagorean theorem to solve right triangles in applied problems.
	SMP:
	• MP2: Reason abstractly and quantitatively.
	• MP4: Model with mathematics.
	• MP7: Look for and make use of structure.

Sample Template 6.1: Continued

Learner Outcomes Note the 21st-century skills taught, practiced, and/or assessed in this project.	**T = Taught; P = Practiced; A = Assessed**	T	P	A	**T = Taught; P = Practiced; A = Assessed**	T	P	A
	Written communication	■	■	□	Technology literacy	■	■	□
	Oral communication	■	■	■	Work ethic	■	■	□
	Collaboration	■	■	■	Civic responsibility	□	□	□
	Critical thinking	■	■	■	Numeracy	■	■	■
	Information literacy	□	□	□	Core content skills	■	■	□

Habits of Mind Indicate one or two habits of mind that are the focus of this unit.	
□ Persisting	□ Thinking about thinking	□ Communicating with clarity	□ Taking responsible risks
□ Managing impulsivity	□ Striving for accuracy and precision	□ Gathering data, using all senses	□ Finding humor
□ Listening to others	■ Questioning, posing problems	■ Creating, imagining, innovating	□ Thinking interdependently
□ Thinking flexibly	□ Applying past knowledge	□ Responding with awe	□ Learning continuously

Presentation Audience

Student Production		
Group products (major types):	• Each group of three or four must agree on a polyhedron and a tiling design for tessellating the faces, then sketch the design, construct the decorated solid, and create a slideshow. • The group may split into mini teams; two people may focus on the design and construction of the solid, while the other two start documenting the measurements, brainstorming ideas for the slideshow presentation, and formatting the slideshow presentation.	**Check all that apply:** ■ Class ■ School □ Community □ Experts □ Web (public) ■ Parents □ Other:
Individual products (major types):	• To help generate design ideas, each student watches different video clips, browses through photo galleries of Archimedean tiling and Islamic drawings, and records their ideas on paper to share with the group. • Each person must record the group-meeting agendas so that everyone tracks the project progression.	

Sample Template 6.1: *Continued*

Assessments and Reflection	**Rubric(s)** Check and describe all that will be used for this project	☐ Multimedia presentation rubric	☐ Other: *Project Rubric*
		■ Oral presentation rubric	☐ Other:
		☐ CCSS ELA and literacy writing rubrics	☐ Other:
		☐ School writing rubric	☐ Other:
		☐ School learner outcomes rubric	☐ Other:
	Assessment Type(s) Check and describe all that will be used for this project	■ Quiz: Written responses in short answer and calculating geometry problems on paper test	■ Performance assessment:
		■ Test: Written responses in short answer and calculating geometry problems on paper test	☐ Notes review:
		☐ Essay:	☐ Checklist:
		☐ Online assessment:	■ Concept map: *Overview of key questions related to the Driving Question*
	Reflection Tools Check and describe all that will be used for this project	■ Survey: Peer and self-evaluation of performance on Google Forms	☐ Focus group:
		☐ Discussion: .	☐ Personal learning plan:
		■ Journal: Written responses to project reflection questions	☐ Student-teacher conference:

Project Resources	On-site personnel:	*Teacher, local artists, art teachers, flooring designers*
	Technology:	*Computers, laptop, calculators, LCD projector, Microsoft PowerPoint Presentation or Google Slides, Paint Program*
	Community resources:	*Faculty from the University of Indianapolis*
	Print resources:	*Blank graph paper, foam boards, cardboard, cardstock paper, white paper, posterboard, tape, glue, rulers, scissors, compasses, markers, crayons, folders*
	Online resources:	*Potential websites for design ideas:* • Annenberg Learner–Interactives: Geometry 3D Shapes (http://www.learner.org/interactives/geometry/3d.html) • Dartmouth College Lesson 5 Art part: Islamic art (http://www.dartmouth.edu/~matc/math5.pattern/lesson5art.html) • Archimedean Tilings (http://www.uwgb.edu/dutchs/symmetry/archtil.htm) • CoolMath4Kids: Tesselations (http://www.coolmath4kids.com/tesspag1.html) Google Forms (http://www.google.com/google-d-s/createforms.html) Wordle (http://www.wordle.net/)

Template adapted from the Buck Institute for Education (www.bie.org).

Sample Template 6.2: *Polyhedron Tiling Artwork* Project Scaffolding NTKs

Scaffolding NTKs: Activity and Assessment Planning

Anticipated Knowledge and Skills Students Need (NTKs)	Assignment or Activity to Address NTKs	How Assignment or Activity Will Be Assessed	Learning Outcomes Addressed in Assignment/ Activity
What is a polyhedron?	• Make geometric solids from 2-D nets in the Entry Event • Use Lesson 1 guided notes to explain characteristics of a polyhedron	• Quiz on Polyhedra and Surface Area • Warm-Up Questions • Geometric Solids Test	• Students will select a polyhedron to build and describe the characteristics of their solid in a Project Planner.
What is surface area?	• Make geometric solids from 2-D nets in the Entry Event • Complete an analysis assignment on the faces of the geometric solids • Use Lesson 2 guided notes to derive surface area formula	• Math Lab: Painting a House • Math Lab: More Applications on Surface Area • Quiz on Polyhedra and Surface Area • Warm-Up Questions • Compute the surface area of the team's chosen polyhedron • Geometric Solids Test	• Students can see how 2-D nets of geometric solid model the surface area of a solid. • Students can see the application of surface area to putting up a tent, painting a house, and various other real-life situations.
What is volume, and how does it differ from surface area? ?	• Use Lesson 3 and 4 to derive volume formulas and solve various example problems	• Math Lab: Surface Area vs. Volume • Warm-Up Questions • Compute the volume of the team's chosen polyhedron • Geometric Solids Test	Students can see that volume means how much space a geometric solid occupies in 3-D space. They can distinguish the difference between surface area and volume.
How do we calculate the area of any regular polygon?	• Use Lesson 5 to show students how to calculate apothems and apply apothems in area calculations	• Students find the volume and surface area of a pentagonal prism • Warm-Up Questions • Geometric Solids Test	Students can see how trigonometry is related to calculating the area of regular polygons.
What is a similar solid?	• Use Lesson 6 to show students how the Similar Solids Theorem is derived • Do various examples using guided notes	• Practice Test Questions • Warm-Up Questions • Geometric Solids Test	• Practice Test Questions • Warm-Up Questions • Geometric Solids Test

Template adapted from the Buck Institute for Education (www.bie.org).

Sample Template 6.3: Polyhedron Tiling Artwork Project Six Problem-Solving Phases Planning Worksheet

PBL Process Phase	NTKs	Scaffolding	Before Moving to Next Phase . . .
Phase 1 What is the need or problem?	Students will need guided notes to help them understand the derivation of the volume and surface area formulas. They will create a polyhedron to demonstrate their understanding of geometric solids.	Students will complete short math labs and practice their mathematical literacy by choosing words from a word bank to explain their answers. They will present their final product and complete a summative assessment.	Students can explain the difference between surface area and volume with a 2-D net of a polyhedron.
Phase 2 What needs to be in our solution?	Students will need to understand the definition of a tiling pattern, characteristics of a polyhedron, and surface area and volume specific to their solids. They need to know how to present their final product, using appropriate academic language, and how to speak confidently about their work.	Students will plan their solid design and tiling pattern using a Project Planner, and then implement their design when they make their 3-D solids. They will present their final product to an authentic audience, who will look for mathematical academic language in the formal presentations.	Students make notecards of what they will say in their final presentation, with key mathematical vocabulary highlighted.
Phase 3 What are possible solutions?	Students will need ideas from accredited websites. After they look at various tiling patterns in Archimedean and Islamic tiling artwork, they will be given time to brainstorm ideas with their team members. They will share ideas on Google drives.	Students can create multiple tiling patterns for their decorated solids. They can also build their solids using their own method after the teacher approves the idea.	Students have digital files of their tiling patterns in a Google drive that is shared with the teacher and team members.
Phase 4 Which solution should we use?	Team members will vote on the best structure and materials that fit their design. They will complete a Project Planner to help them narrow down their ideas.	Students will build the solid based on the design features in the Project Planner. They will explain their work in a dress rehearsal for their formal presentation.	1. Team members score one another's presentations, using the project rubric. 2. Written feedback is given immediately to all teams.
Phase 5 How do we create, run, and inspect our solution?	Students will need lessons on surface area and volume concepts. They will need to break down the process of geometric solid construction into smaller tasks.	Students will present their implementations and results to an authentic audience in a formal presentation that requires them to use math terminology and show their methods for volume and surface area calculations.	1. The judging panelists score each team's presentation and solid, using the project rubric. 2. Peers score every presenting team's presentation and solid, using the project rubric.

Template adapted from the Buck Institute for Education (www.bie.org).

Sample Template 6.4: *Polyhedron Tiling Artwork Project Calendar*

Project: Polyhedron Tiling Artwork Project			Start Date: After Covering Right-Triangle Trigonometry	
Monday	**Tuesday**	**Wednesday**	**Thursday**	**Friday**
		WEEK 1		
1	**2**	**3**	**4**	**5**
Assign teams Entry Event Lesson 1: Polyhedra and Euler's theorem Question-and-answer session about project rubric Goals: • Students will become familiar with their teammates' skills and interests • Students will know how to identify polyhedrons and apply Euler's theorem • Students will understand the grading criteria in the rubric	Lesson 2: Surface Area of Prisms, Cylinders, Pyramids, and Cones View websites about Archimedean tiling and Islamic artwork Introduce Polyhedron Tiling Artwork Project Planner and Time Log Documentation Sheet Goals: • Students will learn about surface area • Students will brainstorm ideas about their decorated solid • Students will sign team contracts in the Project Planner	Lesson 2: Surface Area of Prisms, Cylinders, Pyramids, and Cones Math lab on surface area Project Planner work time Update Time Log Documentation Sheet Goals: • Students will apply surface area in application problems • Students will select their solids and begin creating the tiling pattern • Each member will be assigned a specific task	Go over surface area of prisms, cylinders, pyramids, and cones, as needed Formative quiz on surface area Project Planner work time Update Time Log Documentation Sheet Due today: Students' rough drafts of Project Planner and Time Log Documentation Sheet Goal: • Students are formatively assessed on surface area	Complete short online activities on Interactive Geometry website Finalize Project Planner Finalize Time Log Documentation Sheet Due today: Students' Project Planner and Time Log Documentation Sheet must be finished Goal: • Seeing different applications of surface area and volume gives students new perspectives
		WEEK 2		
6	**7**	**8**	**9**	**10**
Lesson 3: Volume of Prisms, Cylinders, Pyramids, and Cones, Part 1 More Math Lab activity on surface area Work time for building the polyhedron Goal: • Students are introduced to volume concepts and revisit surface area concepts	Lesson 3: Volume of Prisms, Cylinders, Pyramids, and Cones, Part 2 Introduce Presentation Guidelines Students collect data for the presentation Work time for building the polyhedron Due today: Students must have a skeleton outline of all the presentation slides All faces of the solid must be measured Goal: • Students learn more about volume concepts	Lesson 4: Surface Area and Volume of Spheres Students collect data for the presentation Work time for building the polyhedron Due today: Students' introductory slides must be completed All faces of the solid must be constructed Teams who are behind in polyhedron construction will receive warning notes Goal: • Students are introduced to volume and surface area of non polyhedrons	Math lab on surface area and volume Students collect data for the presentation Work time for building the polyhedron Due today: Student's body slides of the presentation must be completed The entire solid must be 75% assembled Goal: • Students review surface area and volume, as necessary	Hand out practice test, and review questions on surface area and volume Students collect data for the presentation Work time for building the polyhedron Due today: The rough draft of the presentation must be completed The solid must be completed Goal: • Students receive a study guide for the summative assessment

Sample Template 6.4: Continued

Monday	Tuesday	Wednesday	Thursday	Friday
		WEEK 3		
11	**12**	**13**	**14**	**15**
Lesson 5: Apothems Work on practice test Go over questions on practice test Reteach any lessons (if necessary) *Goals:* • *Students learn how to calculate apothems and area of regular polygons* • *Students review content knowledge for the formative assessment*	Lesson 6: Similar Solids Work on practice test Go over questions on practice test *Goals:* • *Students are introduced to the Similar Solids theorem* • *Students review content knowledge for the formative assessment*	Dress rehearsal of presentations Give feedback re: teams' presentations Students grade peers in the dress rehearsal and provide constructive feedback *Goals:* • *Students practice public speaking* • *Students have an opportunity to clarify any math misconceptions* • *Students speak about the mathematics they constructed* • *Students do peer evaluations*	Go over last-minute questions re: practice test Geometric Solids Test Day! *Goal:* • *Students' content knowledge of geometric solids is assessed*	Presentation Day! Students grade peers' presentations and provide constructive feedback Complete self-, team, and project evaluations *Goals:* • *Students present their final product to an authentic audience* • *Students reflect on and provide feedback about their PBL journey*

Sample Template 6.5: *Polyhedron Tiling Artwork Project*
Entry Event

Name_____ Date_____ Period_____ MWF/TTH

Getting Started: Exploring Polyhedrons

Complete the table below for each solid. Use the 2-D net cutouts to complete the table.

Draw the 2-D view of the polyhedron.	Draw the 3-D view of the polyhedron.	
What is the name of the polyhedron?	Answer:	
Name and draw all the polygons that form the faces of the solid.	Answer:	
Total vertices:		Total edges:		Total faces:	

Draw the 2-D view of the polyhedron.	Draw the 3-D view of the polyhedron.	
What is the name of the polyhedron?	Answer:	
Name and draw all the polygons that form the faces of the solid.	Answer:	
Total vertices:		Total edges:		Total faces:	

Sample Template 6.5: *Continued*

Draw the 2-D view of the polyhedron.	Draw the 3-D view of the polyhedron.
What is the name of the polyhedron?	Answer:
Name and draw all the polygons that form the faces of the solid.	Answer:

Total vertices:		Total edges:		Total faces:	

Draw the 2-D view of the polyhedron.	Draw the 3-D view of the polyhedron.
What is the name of the polyhedron?	Answer:
Name and draw all the polygons that form the faces of the solid.	Answer:

Total vertices:		Total edges:		Total faces:	

In your own words, what do you think *surface area* means?

What do you think *volume* means?

Sample Template 6.5: *Continued*

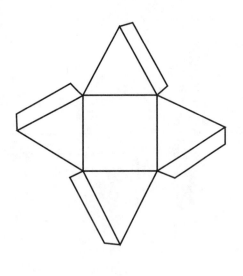

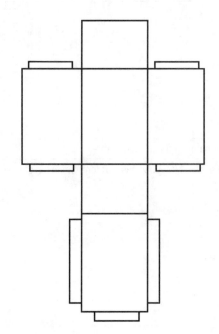

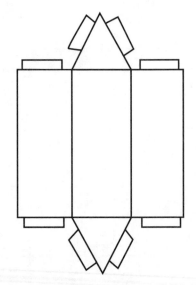

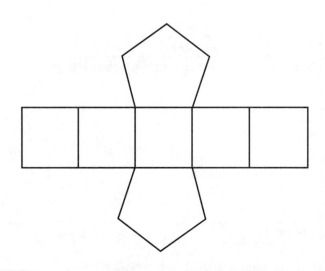

Sample Template 6.6: *Polyhedron Tiling Artwork Project Rubrics*

Group Names _____   Due Date _____   Class _____

For each bullet point met, a box will be checked. Each checked box is equivalent to 1 point.

CRITERIA	EXEMPLARY Surpasses Excellent Requirements	EXCELLENT Surpasses Basic Requirements	BASIC Meets Most Requirements
Artwork (35%) **Number of Boxes Checked** _____	All excellent requirements PLUS: [ ] Uses at least 4 polygons for tiling pattern. [ ] Includes at least 7 repetitions in tiling. [ ] Polyhedron has at least 6 faces. [ ] Polyhedron has at least 2 different types of polygons in its faces. [ ] Applies more than 3 types of color in decorating.	All basic requirements PLUS: [ ] Uses at least 3 polygons for tiling pattern. [ ] Includes at least 6 repetitions in tiling. [ ] Tiling fits perfectly onto polyhedron without truncating any patterns. [ ] Polyhedron has at least 5 faces. [ ] Applies more than 2 types of color in decorating.	[ ] Uses at least 3 polygons for tiling pattern. [ ] Includes at least 5 repetitions in tiling. [ ] Has no gaps or overlaps to fill in the tiling space. [ ] Uses a polyhedron to display tiling. [ ] Applies 2 colors in decorating. [ ] Work is neat. [ ] Project Planner is successfully completed.
Mathematical Content (45%) **Number of Boxes Checked** _____	All excellent requirements PLUS: [ ] Gives informal arguments for the formulas for the volume of a prism, cylinder, pyramid, and cone and uses them to solve problems. [ ] Identifies the shapes of two-dimensional cross-sections of three-dimensional objects, and identifies three-dimensional objects generated by rotations of two-dimensional shapes. [ ] Uses geometric shapes, their measures, and their properties to describe objects, and applies geometric methods to solve design problems. [ ] Entry Event activity is completed with at least 90% accuracy. [ ] All daily warm-ups are completed with at least 90% accuracy. [ ] Individual test score of at least 90% accuracy. [ ] Individual quiz score of at least 90% accuracy.	All basic requirements PLUS: [ ] Gives informal arguments for the formulas for the volume of a prism, cylinder, pyramid, and cone and uses some of them to solve problems. [ ] Identifies the shapes of two-dimensional cross-sections of three-dimensional objects, and identifies three-dimensional objects generated by rotations of some two-dimensional shapes. [ ] Uses geometric shapes, their measures, and some of their properties to describe objects. [ ] Uses trigonometric ratios and the Pythagorean Theorem to solve right triangles in applied problems. [ ] Entry Event activity is completed with at least 80% accuracy. [ ] All daily warm-ups are completed with at least 80% accuracy. [ ] Individual test score of at least 80% accuracy. [ ] Individual quiz score of at least 80% accuracy.	[ ] Gives informal arguments for some of the formulas for the volume of a prism, cylinder, pyramid, or cone. [ ] Identifies the shapes of two-dimensional cross-sections of some three-dimensional objects. [ ] Uses geometric shapes to describe objects. [ ] Constructs an equilateral triangle, a square, and a regular hexagon inscribed in a circle. [ ] Reflective journals are fully completed. [ ] Entry Event activity is completed with at least 70% accuracy. [ ] All daily warm-ups are completed with at least 70% accuracy. [ ] Individual test score of at least 70% accuracy. [ ] Individual quiz score of at least 70% accuracy.
Ms. So-and-So Art Presentation (20%) **Number of Boxes Checked** _____	All excellent requirements PLUS: [ ] Team speaks with enthusiasm. [ ] Slideshow demonstrates some historical or real-world application of the mathematical content. [ ] Team shares skills that they learned or ideas for project improvement.	All basic requirements PLUS: [ ] There is an even distribution of speaking among team members. [ ] Team maintains eye contact to the audience. [ ] Information is confidently presented. [ ] Slideshow has a balance of text and pictures.	[ ] Team speaks audibly to the audience. [ ] Team shows professionalism. [ ] Team shows some eye contact. [ ] Final solid is decorated and displayed. [ ] Discusses faces, vertices, and edges. [ ] Discusses the volume of the solid. [ ] Discusses the surface area of the solid. [ ] Uses at least 20 math vocabulary words.

Final Project Grade: _____%(35) + _____%(45) + _____%(20) = _____

Sample Template 6.6: *Continued*

Group Names _____ Due Date _____ Class _____

Polyhedron Tiling Artwork Judging Rubric for Presentation Day

Directions: For each requirement met, please check the box. Please add comments in the space provided.

CRITERIA	EXEMPLARY Surpasses Excellent Requirements	EXCELLENT Surpasses Basic Requirements	BASIC Meets Most Requirements
Ms. So-and-So Art Presentation	All excellent requirements PLUS: [ ] Team speaks with enthusiasm. [ ] Slideshow demonstrates some historical or real-world application of the mathematical content. [ ] Team shares skills that they learned or ideas for project improvement. TOTAL BOXES CHECKED: _____	All basic requirements PLUS: [ ] There is an even distribution of speaking among team members. [ ] Maintains eye contact with the audience. [ ] Information is confidently presented. [ ] Slideshow has a balance of text and pictures. TOTAL BOXES CHECKED: _____	[ ] Team speaks audibly to the audience. [ ] Team shows professional presence. [ ] Team shows some eye contact. [ ] Final solid is decorated and displayed. [ ] Discusses faces, vertices, and edges. [ ] Discusses the volume of the solid. [ ] Discusses the surface area of the solid. [ ] Uses at least 20 math vocabulary words. TOTAL BOXES CHECKED: _____

COMMENTS

Artwork:

Mathematical Content:

Presentation Skills:

Teamwork:

Final Presentation Grade: Enter Total Boxes Checked in All Categories Here: _____ /15

Sample Template 6.7: *Polyhedron Tiling Artwork Project Team Roles*

Each team member will dedicate his or her hard work to this project. Each member will also be responsible for certain specific tasks.

Group Name _____

Team Leader
Tracks the project's progression

_____

Task Manager
Keeps the team on task and manages time

_____

Technician
Records all the team's work in digital files

_____

Researcher
Finds project ideas and checks for content accuracy

_____

The team's portfolio must remain in the classroom at all times!

Sample Template 6.8: *Polyhedron Tiling Artwork Project* Time Log Documentation Sheet

Building the Solid		
Selecting and sketching the 2-D net of the solid	Time devoted:	Names of members contributing to this task:
Measuring and cutting materials to construct the solid	Time devoted:	Names of members contributing to this task:

Creating the Tiling Pattern		
Coming up with the tiling pattern	Time devoted:	Names of members contributing to this task:
Coloring and placing the tiling patterns on the solid	Time devoted:	Names of members contributing to this task:

Making the Slideshow		
Organizing the content of the slideshow	Time devoted:	Names of members contributing to this task:
Editing the slideshow	Time devoted:	Names of members contributing to this task:

Mathematical Calculations		
Calculate the faces, vertices, and edges of the polyhedron	Time devoted:	Names of members contributing to this task:
Calculating the volume of the polyhedron	Time devoted:	Names of members contributing to this task:
Calculating the surface area of the polyhedron	Time devoted:	Names of members contributing to this task:
Calculating the area of each tiling pattern	Time devoted:	Names of members contributing to this task:

Sample Template 6.9: *Polyhedron Tiling Artwork* Project Planner

Name_____ Date_____ Class _____

Due Date _____

- Your team may not begin the polyhedron construction until the teacher approves this Project Planner.
- Each team member must fill out his or her own copy of the Project Planner.

Directions

Choose one polyhedron from the list below that your team will build:

| Rectangular prism | Octahedron | Square pyramid |
| Triangular prism | Hexagonal prism | Pentagonal prism |

Other (this must be a polyhedron approved by the teacher):

_____

2. Draw a 2-D net of your solid below.

3. What polygons are in the faces of the selected polyhedron?

_____

4. What is the number of vertices, edges, and faces in your polyhedron? Use Euler's formula to assist you in the calculations.

Sample Template 6.9: *Continued*

5. Draw each type of face once, with the appropriate dimensions labeled.

6. List the polygons that you will use to make your tiling pattern. Draw your tiling pattern on a separate document by hand or on the computer (in a Word document or using a paint program).

7. Calculate the volume of your solid.

8. Calculate the surface area of your solid.

Sample Template 6.9: *Continued*

9. Calculate the area of one tiling pattern.

10. How can you find out how many times your tiling pattern needs to be repeated to cover the entire outer region of your polyhedron? Find the total area occupied by your tiling pattern repetitions.

11. Make a list of materials you will need to build your decorated polyhedron.	12. Identify the team members who will be responsible for obtaining the materials. Remember: Everyone must contribute.

Here are some suggestions for constructing your decorated polyhedron:

- Make copies of the tiling pattern in a Word document or a painting program.
- Print or duplicate the tiling pattern by hand, and then paste them onto the solid.
- Cut out the shapes of the tiling pattern, and glue them onto your solid.

Sample Template 6.9: *Continued*

13. Refer to the Time Log Documentation Sheet. Make an agenda of what each person will do to construct the decorated solid.

Team Member 1 Name:	What will this member contribute?
Please have Team Member 1 sign his or her name here	X
Team Member 2 Name:	What will this member contribute?
Please have Team Member 2 sign his or her name here	X
Team Member 3 Name:	What will this member contribute?
Please have Team Member 3 sign his or her name here	X
Team Member 4 Name:	What will this member contribute?
Please have Team Member 4 sign his or her name here	X

Sample Template 6.10: *Polyhedron Tiling Artwork Project* Lesson Plans

Lesson Plans: High School Geometry	
DAY 1: Launching the Project	
Overview Students are assigned to teams. The project is launched with the Entry Event and the project rubric. Lesson 1: Polyhedra and Euler's theorem are introduced. Class ends with a brief question-and-answer session. TOTAL TIME: 60 minutes	**Materials** • Getting Started: Exploring Polyhedrons • Scissors • Project Rubric • Lesson 1 Guided Notes
Learning Targets 1. I can name and identify characteristics of a polyhedron. 2. I can apply Euler's theorem to find the number of faces, vertices, and edges of a polyhedron.	**Other Resources** None
Activities 1. Assign students to teams of three or four. (See Sample Template 6.7.) 2. Give students 10–15 minutes to complete the Getting Started: Exploring Polyhedrons activity with their groups. Each team member can be in charge of one solid. (See Sample Template 6.5) 3. Begin Lesson 1 Guided Notes with the students. This will take 20–25 minutes. 4. Give students 5 minutes to read the project rubric. 5. Begin the question-and-answer session on the project rubric. This should take 10–15 minutes.	**Relevanc:** Create team folders for the teams to store all crucial project documents. Record all questions and answers in the question-and-answer session in a live Google document to engage students.

DAY 2: Surface Area	
Overview Lesson 2 on Surface Area of Prisms, Cylinders, Pyramids, and Cones is introduced. Review online resources to give students ideas for tiling artwork. Students begin working on the *Polyhedron Tiling Artwork* Time Log Documentation Sheet and Project Planner. (See Sample Templates 6.8 and 6.9). TOTAL TIME: 60 minutes	**Materials** • Lesson 2 Guided Notes • Internet • Computers • *Polyhedron Tiling Artwork Project Planner* • Time Log Documentation Sheet
Learning Targets 1. I can calculate the surface area of prisms and cylinders. 2. I can identify the polyhedron that my team will create.	

Activities
 	
1. Begin Lesson 2 Guided Notes with the students. This will take 20–25 minutes.
2. In their teams, have students browse through the following websites for 5–8 minutes:
  • Lesson 5 Art part: Islamic art on Dartmouth College's website (http://www.dartmouth.edu/~matc/math5.pattern/lesson5art.html)
  • Archimedean Tilings on the University of Wisconsin Green Bay website (http://www.uwgb.edu/dutchs/symmetry/archtil.htm)
  Allow two computers per team to avoid off-task behavior.
3. Give students 20 minutes to work on the Project Planner—their task is to decide what solid they will create and to fill out page 1. Each student must fill out his or her copy of the Project Planner.
4. Have students read through the tasks on the Time Log Documentation Sheet and discuss them with their team for 5 minutes.
5. Answer any questions from students for the last 5 minutes.
	Rigor:
	
	Use wrapping paper and 2-D nets to explain surface area concepts.

Sample Template 6.10: *Continued*

Lesson Plans: High School Geometry

DAY 3: Surface Area Continued

Overview

Continue Lesson 2: Surface Area of Prisms, Cylinders, Pyramids, and Cones. Students complete the math lab on painting a house. Students plan the dimensions of their solids. The tasks on the Time Log Documentation Sheet must be clear to each team member.

TOTAL TIME: 60 minutes

Materials

- Lesson 2 Guided Notes
- Math Lab: Painting a House
- *Polyhedron Tiling Artwork* Project Planner
- Rulers
- Time Log Documentation Sheet

Learning Targets

1. I can calculate the surface area of pyramids and cones.
2. I can plan the dimensions of my polyhedron for the project.

Activities

1. Finish Lesson 2 Guided Notes with the students. This will take 20–25 minutes.
2. Allow students 15–20 minutes to finish completing Math Lab: Painting a House with their groups.
3. Give students 15 minutes to work on the Project Planner and map out the dimensions of their solids.
4. Give students 5 minutes to finalize the tasks on the Time Log Documentation Sheet and discuss them with their team.
5. Answer any questions from students for the last 5 minutes.

Rigor:

Review the Pythagorean theorem and isosceles triangles with the students before starting on the math lab.

DAY 4: Reteach and Assessment

Overview

Go over Lesson 2: Surface Area of Prisms, Cylinders, Pyramids, and Cones, as needed. Give a short quiz on surface area. Students plan their tiling patterns. The Project Planner must be at least 75% completed.

TOTAL TIME: 60 minutes

Materials

- Lesson 2 Guided Notes
- Quiz on polyhedra and surface area
- Computers
- Internet
- *Polyhedron Tiling Artwork* Project Planner
- Rulers
- Time Log Documentation Sheet

Learning Targets

1. I can show mastery of surface area of prisms, cylinders, pyramids, and cones.
2. I can create a tiling pattern for my team's solid.

Activities

1. Reteach Lesson 2 Guided Notes with the students. This will take about 15 minutes.
2. Allow students 15–20 minutes to complete the quiz on polyhedra and surface area.
3. Give students 20 minutes to work on the Project Planner and create their tiling pattern.
4. Allow students to use computers to get ideas and draw their tiling pattern.
5. Answer any questions from students for the last 5 minutes.

Rigor:

Push students to draw sketches and show all their work in the quiz.

Encourage students to employ software—such as Microsoft Paint, Microsoft Word, or PowerPoint—to create their tiling patterns.

Sample Template 6.10: *Continued*

Lesson Plans: High School Geometry

DAY 5: Finalize the Plan

Overview	**Materials**
Students complete the short online Interactive Geometry activities. Students finalize their Project Planners and Time Log Documentation Sheets. TOTAL TIME: 60 minutes	• Computers • Internet • *Polyhedron Tiling Artwork Project* Planner • Rulers • Time Log Documentation Sheet

Learning Targets	
1. I can complette the structural details of my team's solid. 2. I can create a tiling pattern for my team's solid.	

Activities	**Rigor:**
1. Give students 30 minutes to complete the 3-D Shapes section of the Interactive Geometry website (http://www.learner.org/interactives/geometry/3d.html). 2. Give students 25 minutes to finalize the entire Project Planner, the solid design, and the tiling pattern. 3. Allow students to use computers to get ideas and draw their tiling pattern. 4. Have students complete the Time Log Documentation Sheet. 5. Answer any questions from students for the last 5 minutes.	The Interactive Geometry activity helps students explore four prisms. Have students work in pairs at computers to promote dialogue. Students can change the direction and speed of the prisms' rotation and explore the corresponding nets of each prism.

DAY 6: Volume

Overview	**Materials**
Introduce Lesson 3: Volume of Prisms, Cylinders, Pyramids, and Cones. Students complete another math lab on surface area. Students gather materials for their solids and measure and cut out the faces of their solids. TOTAL TIME: 60 minutes	• Lesson 3 Guided Notes • Math Lab: More Applications on Surface Area • Time Log Documentation Sheet • *Polyhedron Tiling Artwork* Project Planner • Computers • Internet • Rulers, scissors • Tape, glue, stapler • Construction paper • Cardstock • Blank paper

Learning Targets	
1. I can calculate the volume of prisms and cylinders. 2. I can calculate the surface area of right prisms. 3. I can gather materials for my team's solid and create the 2-D net of the solid.	

Activities	
1. Begin Lesson 3 Guided Notes with the students. This will take about 20 minutes. 2. Allow students 20 minutes to complete the Math Lab: More Applications on Surface Area with their teams. 3. Give students the last 20 minutes of class to plan the materials they will need to construct their polyhedra and to measure the actual materials they need for the construction process. One person from each team will record the materials and measurements in the Project Planner, and one person will gather construction materials.	

Sample Template 6.10: *Continued*

Lesson Plans: High School Geometry

DAY 7: Volume Continued

Overview

Continue Lesson 3: Volume of Prisms, Cylinders, Pyramids, and Cones. Explain the requirements in the Ms. So-and-So Art Presentation Guidelines. Students start an outline of their presentation slideshow. Students continue to gather materials for their solids and to measure and cut out the faces of their solids.

TOTAL TIME: 60 minutes

Learning Targets

1. I can calculate the volume of pyramids and cones.
2. I can make an outline for my project presentation.
3. I can begin assembling the faces of my team's solid and embed the tiling pattern onto the faces.

Materials

- Lesson 3 Guided Notes
- Ms. So-and-So Art Presentation Guidelines
- Mathematical Literacy handout
- Time Log Documentation Sheet
- *Polyhedron Tiling Artwork* Project Planner
- Computers
- Internet
- Rulers, scissors
- Tape, glue, stapler
- Construction paper
- Cardstock
- Blank paper

Activities

1. Finish Lesson 3 Guided Notes with the students. This will take about 20 minutes.
2. Use 10 minutes to explain the Ms. So-and-So Art Presentation Guidelines and the Mathematical Literacy handout.
3. Give students the last 30 minutes of class to create an outline of their presentation, assemble the faces of their solids, and embed the tiling patterns onto the solid.

Relationship:

Have students save their presentation file in their Google account and share the file with the teacher and their group members. Doing so will maximize the opportunities to access and revise the team's work.

DAY 8: Spheres

Overview

Introduce Lesson 4: Surface Area and Volume of Spheres. Students need to input the dimensions of their solids into their slideshows and select the 20 words they will include in their presentations. Students begin creating their slideshows. Students paste their tiling patterns onto their solids.

TOTAL TIME: 60 minutes

Learning Targets

1. I can calculate the volume and surface area of spheres
2. I can make the introductory slides of my presentation.
3. I can finish assembling the faces of my team's solid and embed the tiling pattern onto the faces.

Materials

- Lesson 4 Guided Notes
- Ms. So-and-So Art Presentation Guidelines
- Mathematical Literacy handout
- Time Log Documentation Sheet
- *Polyhedron Tiling Artwork* Project Planner
- Computers
- Internet
- Rulers, scissors
- Tape, glue, stapler
- Construction paper
- Cardstock
- Blank paper

Activities

1. Begin Lesson 4 Guided Notes with the students. This will take about 25 minutes.
2. The remainder of the class is project work time. Students must complete the introductory slides for their slideshow. The decorated faces of the solid must be 50% completed

Sample Template 6.10: *Continued*

Lesson Plans: High School Geometry

DAY 9: Math Lab and Work Time

Overview

Students identify the differences between *surface area* and *volume* in a math lab. Students complete the slides for the body of their presentation. Students finish pasting their tiling pattern onto their solids.

TOTAL TIME: 60 minutes

Materials

- Math Lab: Surface Area vs. Volume
- Ms. So-and-So Art Presentation Guidelines
- Mathematical Literacy handout
- Time Log Documentation Sheet
- *Polyhedron Tiling Artwork* Project Planner
- Computers
- Internet
- Rulers, scissors
- Tape, glue, stapler
- Construction paper
- Cardstock
- Blank paper

Learning Targets

1. I can differentiate between *volume* and *surface area*.
2. I can create the slides for the body of my presentation.
3. I can finish assembling the faces of my team's solid and embed the tiling pattern onto the faces.

Activities

1. Give students 30 minutes to complete the Math Lab: Surface Area vs. Volume with their teams.
2. Give teams 10 minutes to brainstorm about the slides for the body of their presentations.
3. Have some team members complete the slides for the body of the presentation, and the remaining team members finish assembling the solids with their tiling patterns.

DAY 10: Practice Test and Finalize

Overview

Review surface area and volume concepts from a practice test. Students must have a rough draft of their presentation completed. Students finish constructing the entire solid.

TOTAL TIME: 60 minutes

Materials

- Geometric Solids Practice Test
- Ms. So & So Art Presentation Guideline
- Mathematical Literacy handout
- Time Log Documentation Sheet
- *Polyhedron Tiling Artwork Project* Planner
- Computers
- Internet
- Rulers, scissors
- Tape, glue, stapler
- Construction paper
- Card stock paper
- Blank paper

Learning Targets

1. I can solve story problems involving surface area and volume.
2. I can finish a rough draft of my team's presentation slides.
3. I can finish building my team's decorated polyhedron.

Activities

1. Give students 10 minutes to work on five problems selected by the teacher from the Geometric Solids Practice Test.
2. Review the five selected problems for 20 minutes.
3. For the remaining 30 minutes, have teams split the work among different members; two members will finish the final touches to the decorated polyhedron, and one or two members will finish the rough draft of the entire slideshow.

Sample Template 6.10: *Continued*

Lesson Plans: High School Geometry

DAY 11: Apothems

Overview Work on Lesson 5: Apothems, and review selected questions from the Geometric Solids Practice Test. Students study for the Geometric Solids Test. Reteach any lessons as needed. TOTAL TIME: 60 minutes	**Materials** • Lesson 5 Guided Notes • Geometric Solids Practice Test • Mathematical Literacy handout
Learning Targets 1. I can find the area of regular polygons by calculating the apothem. 2. I can apply surface area and volume in story problems.	
Activities 1. Teach Lesson 5: Apothems. This will take about 25 minutes. 2. Select five problems from the Geometric Solids Practice Test for students to work on with their teams for 20 minutes. 3. Review the selected five problems from the Geometric Solids Practice Test for the remainder of the class period.	**Rigor:** Review trigonometry during Lesson 5. Select problems on surface area and volume from the Geometric Solids Practice Test for the students to work on.

DAY 12: Similar Solids

Overview Work on Lesson 6: Similar Solids, and review selected questions from the Geometric Solids Practice Test. Students study for the Geometric Solids Test. Reteach any lessons as needed. TOTAL TIME: 60 minutes	**Materials** • Lesson 6 Guided Notes • Geometric Solids Practice Test • Mathematical Literacy handout
Learning Targets 1. I can apply the similar solids theorem to find the scale factor of similar solids and ratios of surface area and volume. 2. I can apply surface area and volume in story problems	
Activities 1. Teach Lesson 6: Similar Solids. This will take about 30 minutes. 2. Select three problems from the Geometric Solids Practice Test for the students to work on with their teams for 15 minutes. 3. Review the selected three problems from the Geometric Solids Practice Test for the remainder of the class period.	**Rigor:** Select problems on surface area and volume from the Geometric Solids Practice Test for students to work on. Remind students about the dress rehearsal of their presentations.

Sample Template 6.10: *Continued*

Lesson Plans: High School Geometry

DAY 13: Dress Rehearsals

Overview	**Materials**
Students practice their presentations and practice presenting solids to the class. They provide constructive feedback to the presenting teams. TOTAL TIME: 60 minutes	• Peer Evaluation for Presentations form • Project Rubric • Ms. So & So Art Presentation Guidelines • Mathematical Literacy handout

Learning Targets	
1. I can use math terminology in my presentation. 2. I can provide constructive feedback regarding my peers' presentations.	

Activities	**Relevance:**
1. Explain to the class how to fill out the Peer Evaluation for Presentation form. 2. Each team gives its presentation. 3. Students observe the presentations and give their completed Peer Evaluation for Presentation forms to the presenting teams. 4. Have teams make a list of revisions to their presentation during the last 10 minutes of class.	Give teams a choice: Each team can choose three to four presentations for which they will complete Peer Evaluation for Presentation forms.

DAY 14: Geometric Solids Test

Overview	**Materials**
Students ask any last-minute questions before the test. Most of the class period is devoted to the Geometric Solids Test. TOTAL TIME: 60 minutes	• Geometric Solids Test • Peer Evaluation for Presentations form • Project Rubric • Ms. So & So Art Presentation Guidelines • Mathematical Literacy handout

Learning Targets	
1. I can show mastery of characteristics of a polyhedron, surface area, volume, and similar solids.	

Activities	**Relationship:**
1. Spend the first 10 minutes of class reviewing any last-minute questions. 2. Give students the remainder of the class period to take the Geometric Solids Test. 3. If students finish the test early, have them revise their presentations based on the Peer Evaluation for Presentation forms as needed.	Have one member from each group send their slideshow to the teacher before the end of the day via email. The teacher can review the slideshow and suggest last-minute revisions, if needed. **Relevance Tip:** Remind students to dress professionally for their presentations, as they would in the real world.

Sample Template 6.10: *Continued*

Lesson Plans: High School Geometry
DAY 15: Final Presentation

Overview Students present their final product and slideshow to an authentic audience. The authentic audience gives constructive feedback after the presentations. Students complete the self-, team member, and project evaluation forms. TOTAL TIME: 60 minutes	**Materials** • Computer • Internet • Projector • Clicker • Peer Evaluation for Presentation form • Mathematical Literacy handout • *Polyhedron Tiling Artwork Project* Judging Rubric • Self-, Team Member, and Project Evaluation forms
Learning Targets 1. I can collaborate with my team to present our decorated polyhedron, using proper mathematical vocabulary, to an authentic audience.	
Activities 1. Give students 5 minutes to prepare for their presentations. 2. Start the presentations. Judges will score each presentation using the *Polyhedron Tiling Artwork Project* Judging Rubric. 3. Have students complete the Peer Evaluation for Presentations form for the presenting teams they selected. 4. After the presentations, allow 10 minutes for the judges to give feedback. 5. Have students take home the Self-, Team Member, and Project Evaluation forms to complete for homework.	**Rigor Tip:** Create Google Forms for the Self-, Team Member, and Project Evaluation forms, so students can truly reflect and evaluate their performance without any pressure.

CHAPTER 7

I'm Hair to Help: A Unit on Philanthropy and Linear Equations

[handwritten: does this isolate certain students?]
[handwritten: is avg. growth reliable data]
[handwritten: work to find places that accept hair from all races]

Julie Evans, *Bloomfield Junior/Senior High School, Indiana*
Enrique Galindo, *Indiana University Bloomington*
Gina Borgioli Yoder, *Indiana University–Purdue University Indianapolis*

Authors' note: The first-person accounts in this chapter are from the perspective of the first author, Mrs. Julie Evans, who was one of approximately forty secondary mathematics teachers from a rural county in a midwestern state participating in professional development funded by a three-year Mathematics Science Partnership (MSP) grant. The other authors served as facilitators of the professional development.

Before creating this PBL unit, the same thing happened each year: I started the chapter on linear functions hoping that my students would retain the concepts taught in the chapter and be able to apply them as we moved into deeper concepts in Algebra 1—and each year, I was disappointed. I felt like a total failure. I thought that I was doing everything I could possibly do to make my students successful, so why did they not remember how to write an equation once we moved on to systems?

It seems so obvious now, but at the time I didn't realize that they were only learning the procedures by rote. They didn't have a true understanding of what they were doing or why those procedures worked. They didn't truly understand what the *y*-intercept and the slope represented, so they were unable to apply those concepts to real-world problems.

As part of an MSP grant, teachers from across my county got together for a week-long professional development workshop focused on teaching and learning mathematics for understanding. In the previous year of the grant, we had learned about PBL and created a unit. This year, three other Algebra 1 teachers and I got together and decided to design a unit on linear functions. We chose this topic because our students always struggle with problems on standardized tests that require them to write linear functions for different contexts.

We immediately started searching popular PBL sites for ideas to get us started and were intrigued by a unit called "Let Down Your Hair," featured on the Mathalicious website (www.mathalicious.com). We liked the humanitarian aspects of this unit and decided to use it as a launch for our own unit.

Standards Addressed

According to the CCSSM, high school students build on the semiformal notions of function they learned in grade 8 and begin to use formal notation and language for functions (Common Core Standards Writing Team 2013). This unit addresses the first three groups of high school function standards: Interpreting Functions (HSF.IF), Building Functions (HSF.BF), and Linear, Quadratic and Exponential Models (HSF.LE). Some number, algebra, and statistics standards and some mathematical practices are also addressed.

This unit challenges students to determine how long it would take them to grow their hair long enough to donate it to a nonprofit organization and benefit someone who is suffering from long-term medical hair loss. Students researched both the regulations on donating hair to the organization and average hair growth rates. While each group of students took a slightly different approach to this problem, all students calculated and interpreted the average rate of hair growth (HSF.IF.B.6, HSN.Q.A.2, and HSF.LE.A.1.B), graphed linear functions to represent hair growth (HSF.IF.C.7.A), wrote functions to describe a relationship between the length of hair and the number of days passed (HSF.BF.A.1 and HSA.CED.A.2), and made a prediction to determine the number of days needed for the hair to be long enough to be donated (HSF.LE.B.5, HSS.ID.B.6.C, and HSS.ID.C.7).

The unit also targeted several 21st-century skills—including oral communication, collaboration, critical thinking, and work ethic—and addressed several SMPs:

- Students worked on this challenge for an extended period of time, thus addressing MP1: Make sense of problems and persevere in solving them.
- In preparing a team presentation to share their findings with their classmates and answer their questions, students addressed MP3: Construct viable arguments and critique the reasoning of others.
- They addressed MP4: Model with mathematics by creating a linear model to represent the situation and mathematically articulating several possible scenarios, including starting hair length, desired hair length, and the impact of a potential haircut to the donation timeline.

See Sample Template 7.4 for a complete list of the CCSSM and SMPs addressed in this project.

Project Highlights

The following lessons highlight how the *I'm Hair to Help* unit is rigorous, is relevant to students, and helps students build relationships with one another. Lesson plans for each highlighted lesson can be found at the end of this chapter.

Day 1: Entry Event and Introduction of the Project

A woman in our community suffers from *alopecia areata*, a medical condition that causes hair loss on part or all of the body. We decided that she would be the perfect guest speaker to get students interested in the Locks of Love organization. Her visit and talk served as our Entry Event.

Our guest spoke with the students about how alopecia has affected her life. She went from being voted the girl with the best hair her senior year in high school to having no hair at all. She talked about the emotional, physical, and social effects of this condition.

She then led the discussion toward a solution to the discomfort and ill effects that she and others like her have experienced. Students immediately brought up the Locks of Love organization, our community partner for the unit. A few had donated their hair before and were able to discuss the requirements based on their real-life experience.

(See Sample Template 7.1 at the end of this chapter for more about the Entry Event and other scaffolding activities throughout the unit.)

Relationships: Before the guest speaker came in, the students and I discussed what to expect. I gave them a brief overview of what alopecia is and asked them to brainstorm some questions we might have for her. Students were to brainstorm at home that night and come prepared with a possible question for her the next day. I would encourage anyone implementing this project to prepare the students ahead of time for the guest speaker, if teachers choose to use one as the Entry Event.

After our speaker left, I posed a question to the class: "How long would it take *me* to be able to donate my hair?" Students were hooked. They started talking about how long my hair was and that it depended on how fast my hair grows, and some even mentioned that it also depended on how long I wanted it to be after it was cut.

I ended Day 1 of the unit by handing out the final rubric for the project (see Sample Template 7.2). Students were told to look it over that evening and to come to class the next day with questions and suggestions for edits. (See Sample Template 7.6 for a more detailed Day 1 lesson plan.)

Day 2: Hair Growth Estimation

Day 2 started with clarifying questions about the rubric. Students were nervous about presenting their final products, so we had a short discussion about who the audience would be and how we would inform and persuade them. We also discussed their options for using technology in their presentations, which ranged from simply scanning their paper-and-pencil work and projecting it for their audience to creating a more professional presentation using Desmos.com or other graphing resources, along with PowerPoint or similar presentation tools.

Rigor: Given a choice, most students opted to simply scan their paper-and-pencil work rather than push themselves to learn new tools. I plan to incorporate the use of graphing tools into future implementations of the unit, as I see them as beneficial to this project. As I use the tools more often in class, my hope is that students will become more comfortable with using them and will therefore be more likely to do so.

I had students recall what they remembered from the discussion the day before about how much time it would take me to grow my hair so that it was long enough to donate. I then had them estimate what an average rate of hair growth might be. Their estimations varied greatly—from 2 inches per month to 5 or 6 inches per year to an inch per week. To assist them with determining a more accurate estimation, I introduced the video of Ryan's hair growth from the Mathalicious website. In the video, Ryan takes his picture each day for several weeks to show the length of his hair. Again, students were hooked. They loved this video and were immediately focused on estimating how much Ryan's hair had grown during each segment.

Students suggested a few different ways of recording the data, ultimately deciding that a graph would be the most useful. After graphing the data, we were able to have a lengthy discussion about what the rate of change might be, what an average rate might be, and what the intercepts of the graph might mean. Using the Mathalicious "Let Down Your Hair" resources, I gave a mini lesson on investigating other graphs, including those with different y-intercepts, zero slopes, and other slopes of all kinds. Students were easily able to interpret the graphs in the context of hair growth and tell me what was happening with the hair of the people in each graph.

Rigor: Students can easily "fall through the cracks" during these mini lessons. Find time during the unit to assess each student individually on these skills. I originally used homework as my assessment, which was somewhat effective, but I now prefer to use an exit ticket/mini quiz. This gives me immediate feedback, and I am able to address any issues the next day when the students return. The important thing is to find an assessment method that works for both the teacher and the students.

I had students work in their groups on the Driving Question for the unit: As philanthropists, how can we determine the amount of time necessary to grow our hair long enough to donate it to Locks of Love?

Each group then chose one student's results to discuss in their final presentation. Students couldn't get the measuring tapes in their hands fast enough!

This class ended too quickly for most of the students; they were so involved with their work that they didn't want to leave! (See Sample Template 7.7 for a more detailed lesson plan.)

Day 3: Linear Equations: Putting It All Together and Presentation Preparation

Day 3 was exciting for me as a teacher. As students worked in their groups to finish their individual predictions, they soon noticed a pattern: They were doing all the same calculations for each person, just with different starting points. I used this as our next mini lesson (see Sample Template 7.3). We were able to write a linear equation in slope-intercept form for each student in the class; and they each understood exactly what the variables, slope, and y-intercept represented. It was a very quick mini lesson, but very beneficial.

We discussed other contexts that could be modeled by linear equations in slope-intercept form, and most students were easily able to write equations for these as well. That's when I really saw the benefits of this project. Writing linear equations is something my students have always struggled with, in class and on standardized tests. I worked with some students longer than others, but I was confident at the end of the unit that all students had mastered writing linear equations in slope-intercept form to represent a relationship between two quantities.

Students' homework was to tie up any loose ends with their final products and to be ready to practice their presentations the next day. (See Sample Template 7.8 for a more detailed lesson plan.)

Relationships: I often struggle with the issue of time. My class periods are only 50 minutes, so trying to fit work time and a mini lesson into one class period is difficult. To minimize these time constraints, I plan to try a new strategy next year based on a tip for managing PBL offered by Andrew Miller (2011). He suggests forming students into teams and assigning them roles, one of which is team leader. The teacher teaches a mini lesson or gives directions only to the team leaders, who then teach their groups. Doing so increases the trust and responsibility between the teacher and the students, as well as between the team leader and his or her peers. The assigned roles allow students to exercise their skills for productive collaboration with both peers and adults.

Day 4: Practice Presentations

Day 4 was rough, to say the least. Students understood the concept of how long it would take them to grow their hair for the purpose of donating it to Locks of Love, and they had created great presentation materials, but they lacked experience and confidence with verbally communicating that information to an audience. We spent the day discussing important features of good presentation and speaking skills.

By the end of the day, each group had given a practice presentation. The student audience scored them on the rubric and provided the presenters with "I like" and "I wonder" comments. This protocol, which is commonly used in PBL, is a great way for students to provide feedback and suggest next steps in a positive way.

Giving students this opportunity to practice before their final presentations had many advantages: Each group got to see others present, they had a chance to experience what it was like to receive feedback and to assess someone else, and they were able to make changes to their presentation based on the feedback they received. An important piece of feedback was that groups should explain the background of this project, which most students had not included. I was so happy that we decided to include this extra day for practicing the presentations. (See Sample Template 7.9 for a more detailed lesson plan.)

Day 5: Presentations

Because of the opportunity to practice, the presentations were great—students were organized, well-spoken, and well-prepared. But beyond that, they had clearly learned the mathematical objectives of the unit.

The audience consisted of me, our principal and assistant principal, the superintendent of our school district, and an 11th-grade English/language arts class and their teacher. I included this class because they had been working on persuasive speeches, and I thought they would be able to provide my students with beneficial feedback on their presentation skills and the quality of their mathematical arguments. I had also invited our guest speaker, but she was unable to attend. Looking back, I think it would also be useful to invite a representative from the Locks of Love organization, our community partner.

Students from each group clearly explained the background of the problem and how they went about solving it. Groups then explained the mathematical process for calculating the amount of time it would take their chosen group member to grow his or her hair long enough to donate to Locks of Love (see the example in fig. 7.1).

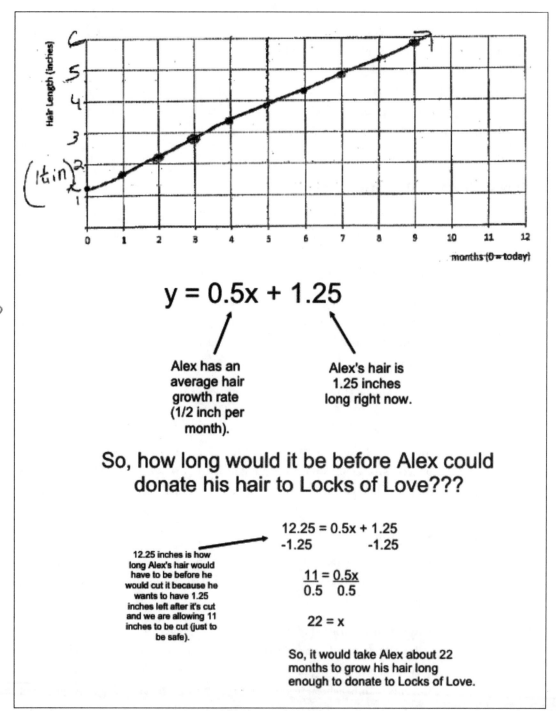

do students factor in how short a student is willing to cut their hair

$$y = 0.5x + 1.25$$

Alex has an average hair growth rate (1/2 inch per month).

Alex's hair is 1.25 inches long right now.

So, how long would it be before Alex could donate his hair to Locks of Love???

12.25 inches is how long Alex's hair would have to be before he would cut it because he wants to have 1.25 inches left after it's cut and we are allowing 11 inches to be cut (just to be safe).

$$12.25 = 0.5x + 1.25$$
$$-1.25 \qquad\qquad -1.25$$

$$\frac{11}{0.5} = \frac{0.5x}{0.5}$$

$$22 = x$$

So, it would take Alex about 22 months to grow his hair long enough to donate to Locks of Love.

Fig. 7.1: Sample student work.

I encouraged the 11th-grade students in our audience to use their mathematical knowledge to question my students about the information they were presenting. This was a great way for me to see how well my students understood the learning objectives.

Most groups ended their presentations with a persuasive piece in which they endeavored to convey the importance of donating to the Locks of Love organization. I was blown away by their passion in doing so. Every group had a story to tell about an aunt, a parent, a friend who had undergone chemotherapy or suffered from other medical conditions that caused them to lose their hair. Their stories were persuasive and personable, and they gave the audience a reason to donate hair, time, or even money to the Locks of Love organization. In fact, at the end of the unit, Diane Rodriguez (one of the four project designers) and a student from this class donated their hair to Locks of Love! (See Sample Template 7.10 for a more detailed lesson plan.)

Addressing the Six A's

When developing this project, the teachers kept in mind the Six A's for designing PBL projects (Markham, Larmer, and Ravitz 2003):

- Both the guest speaker and the context of the Locks of Love organization gave the unit a strong sense of authenticity. Meeting someone who could benefit from this organization helped make the problem meaningful and relevant to the students—they saw it as a question that adults in the real world might tackle.
- The guest speaker and the audience for the presentations, which included other teachers and administrators, provided adult connections.
- The unit allowed students to demonstrate applied learning: They worked in teams, used technology, communicated their ideas, and applied mathematical knowledge to a realistic and complex problem.
- Academic rigor was addressed in that the Driving Question posed a challenge that allowed students to meet important CCSSM.

The CCSSM (National Governors Association Center for Best Practices and Council of Chief State School Officers 2012) call for three main shifts in mathematics: Focus, coherence, and rigor. *Rigor* is defined as pursuing conceptual understanding, procedural skill and fluency, and application with equal intensity. The teachers were motivated to design this unit because they had experienced some shortcomings when teaching the topic of linear function in the past. They realized that their students did not have a true understanding of the topic and the procedures they were using. The PBL unit targeted conceptual understanding and application by having students model a real situation and use different mathematical representations to do so. This unit design also addressed mathematical rigor: Students learned different mathematical strategies from their classmates as groups presented their solutions.

Mrs. Evans describes how the unit also supported students in developing personal relationships:

> The relationships the teacher has with the students, and the relationships the students have with one another, can really set the tone of the class for the entire year. I have found that because of this unit, my students feel more comfortable opening up to one another. They

do not hesitate to ask questions in class, and they do not hesitate to help one another when they need it. Students also formed relationships with the guest speaker and their audience members, simply because they shared their personal stories. The five days spent on this unit saved countless hours of re-explaining concepts because a student was too afraid to ask for help during the initial explanation.

21st-Century Workplace Skills

Cathy Seeley (2013), former president of the National Council of Teachers of Mathematics, said, "All students, whether they are headed toward STEM majors, business majors, other majors, or career paths not calling for four-year degrees, need to be prepared with workplace skills" (p. 16). The Buck Institute for Education (Markham, Larmer, and Ravitz 2003) also acknowledges the importance of 21st century workplace skills:

> Effective workers can productively utilize resources, interpersonal skills, information, systems and technology. Additionally, they demonstrate competency in basic skills in reading, writing, mathematics, speaking and listening; in thinking skills such as creativity, making decisions and solving problems; and in personal qualities such as individual responsibility, self-esteem, self-management, sociability and integrity. (p. 26)

Keeping students' college and career readiness in mind, Mrs. Evans intentionally embedded a number of 21st-century workplace skills into this unit:

* Collaboration and decision-making
* Learning and applying social skills to navigate group interactions
* Using technology for learning and communicating
* Investigating dilemmas using problem solving and critical thinking skills
* Using communication skills to present information generated through investigation, research, and reasoning
* Developing an understanding of and empathy for another (ethics)
* Working together to take action regarding a social dilemma (civic responsibility)

Conclusion

In the professional development provided by the MSP, we learned about criteria for selecting good tasks. This unit gave us an opportunity to move beyond simply selecting good mathematical tasks to actually designing a project-based unit.

It is clear that this unit deeply engaged the students. I was so excited to see them jump into this project with eagerness. They were tired of learning (or not learning) math through lectures and procedures, and I saw a spark in them I had never seen before! Students who always seemed to struggle were suddenly excelling—they were helping "the smart kids" write their equations and do their calculations. I didn't have to circulate the room for the purpose of tapping the sleepers and the doodlers on the shoulders. Instead, I was able to circulate the room to hear the thoughts of my students and to question their thinking. Isn't that how teaching math should always be?

I have taught this unit for two years now, and I am happy to say that I am no longer disappointed at the end of my linear functions unit. Each year, I'm thrilled to see that my students are able to write linear functions, interpret a rate of change and intercepts, and identify other linear models in context. They retain all of this material and can easily apply it to later concepts in Algebra 1, such as systems. This unit also taught students how to present material in a clear, organized way.

But I am also thrilled with the life lesson the students come away with after this unit. They have a new understanding and empathy for people with medical conditions such as alopecia. A lot of students were able to reflect on their own experiences with family members and friends who had undergone chemotherapy during cancer. One student spoke for the first time about his mother, who had died of breast cancer—this unit was a great emotional outlet for him. Those types of lessons are greater than any math lesson I could ever teach them.

References

Common Core Standards Writing Team. *Progressions for the Common Core State Standards in Mathematics* (draft). *Grade 8, High School, Functions.* Tucson, Ariz.: Institute for Mathematics and Education, University of Arizona, 2013.

Markham, Thom, John Larmer, and Jason Ravitz. *Project Based Learning Handbook: A Guide to Standards-Focused Project Based Learning.* 2nd ed. Novato, Calif.: Buck Institute for Education, 2003.

Mathalicious. *Let Down Your Hair,* July 2014, http://www.mathalicious.com/lessons/let-down-your-hair.

Miller, Andrew. "Twenty Tips for Managing Project-Based Learning." *Edutopia*, September 14, 2011, http://www.edutopia.org/blog/20-tips-pbl-project-based-learning-educators-andrew-miller.

National Governors Association Center for Best Practices and Council of Chief State School Officers. *Common Core State Standards Mathematics.* Washington, D.C.: Author, 2012.

Seeley, Cathy. "Learning to Work." *National Council of Supervisors of Mathematics Newsletter* 44 (Winter 2013): 16, http://www.mathedleadership.org/resources/newslettersvol44.html.

Sample Template 7.1: *I'm Hair to Help* Scaffolding NTKs

Scaffolding NTKs: Activity and Assessment Planning

Anticipated Knowledge and Skills Students Need (NTKs)	Assignment or Activity to Address NTKs	How Assignment or Activity Will Be Assessed	Learning Outcomes Addressed in Assignment/Activity
What is Locks of Love and why would someone need it?	Guest speaker and discussion	Persuasive and informative presentation at end of unit	Students should have an understanding of what the guest speaker has had to deal with (medically, emotionally, socially, & cosmetically).
What is an average hair growth rate?	Discussion and video of Ryan's hair growth over a year	• Participation in the discussion • Application of average hair growth rate in final products	Students should be able to find the average for hair growth to predict how long it will take for their hair to grow the required 10 inches for Locks of Love.
What do the slope and intercepts of the graph represent?	Mini lesson	Students are asked to interpret graphs representing hair length over time	Students should be able to interpret the meaning of the slope and intercepts of a graph with regard to a context.
How do I write a linear equation to represent the context?	Mini lesson and discussion	End-of-unit presentation and quiz	Students should be able to write a linear equation to represent a context.
How do I present my findings in an organized and effective way?	Practice presentations with feedback	Final presentations at end of unit	Students should be able to design a multimedia persuasive presentation involving all group members.

Template adapted from the Buck Institute for Education (www.bie.org).

Sample Template 7.2: *I'm Hair to Help* Project Rubric

Criteria	Failing Frizzies (Below Performance Standards)	So-So Strands (Minimal Criteria)	Hair-Raising Heroes (Demonstrates Exceptional Performance)
Math Reasoning (30%)	• Is able to recognize linear equations • Fails to create a linear equation from the given data • Can solve the equation, but cannot communicate how this relates to an individual person's donating timeline 0 - - - - - - - - 9 - - - - - - - - 18	• Creates a linear equation from the given data • Can solve the equation and somewhat communicate how this relates to an individual person's donating timeline • Can articulate, mathematically, the importance of starting hair length 19 - - - - - - - - 22 - - - - - - - - 26	*In addition to meeting the ACCEPTABLE (So-So Strands) criteria:* • Exhibits the breadth and depth of specific knowledge of central concepts required by this problem • Can articulate, mathematically, the importance of desired ending hair length • Discusses the impact of a potential haircut on donation timeline 27 - - - - - - - - 30
Work Ethic and Collaboration (20%)	• Completes assignments only with encouragement • Completes assignments when class time is available, but does not do work outside of class • Is unengaged and unproductive during class discussions • Submits assignments after deadlines 0 - - - - - - - - 6 - - - - - - - - 12	• Comes to class prepared • Contributes meaningfully to small-group discussions • Completes in-class assignment with steady focus • Does not create distractions for other students • Participates in group presentation • Completes out-of-class assignments with accuracy and detail • Completes peer and self-evaluations 13 - - - - - - - - 15 - - - - - - - - 17	*In addition to meeting the ACCEPTABLE (So-So Strands) criteria:* • Exhibits leadership, regardless of project role • Seeks constructive feedback prior to project deadlines • Makes adjustments to improve project performance as recommended by instructor 18 - - - - - - - - 19 - - - - - - - - 20

Final Presentation (50%)	• Body language:
	  – Facial expressions distract from presentation
	  – Not enthusiastic
	  – Eye contact not maintained
	• Speech:
	  – Volume too loud/too soft for setting
	  – Words are not enunciated
	  – Pace of speech is difficult to understand
	  – Uses filler words (uh, um, like)
	• Presentation is less than five minutes
	• Not all members participate
	• Presentation is mostly data and is nonpersuasive
	• Members do not answer the audience's questions
	• Members depend mostly on notes
	
	0 - - - - - - - - 15 - - - - - - - - 30
		• Body language:
		  – Uses some enthusiasm (smile, inflection)
		  – Eye contact maintained
		• Speech:
		  – Volume is appropriate for setting
		  – Words are enunciated, and speech is understandable
		  – Uses minimal filler words (uh, um, like)
		• Presentation is at least five minutes
		• All members participate in some capacity
		• Presentation is persuasive and uses data but not cohesively
		• Members mostly answer the audience's questions
		• Members use some notes
		
		31 - - - - - - - - 37 - - - - - - - - 44
			In addition to meeting the ACCEPTABLE (So-So Strands) criteria:
			• Body language:
			  – Gestures, stance, and expressions enhance presentation
			  – Talks directly to audience members who ask questions, as well as pulls others in
			• Speech:
			  – Has enthusiasm and clarity
			  – Avoids filler words (uh, um, like)
			• All members participate equally
			• Members do not read from notes
			• Presentation uses data to persuade, information is fluid, and transitions from data to persuasiveness are smooth and well-rehearsed
			• Members answer the audience's questions completely and correctly (mathematically)
			
			45 - - - - - - - - 47 - - - - - - - - 50

Sample Template 7.3: *I'm Hair to Help* Mini Lesson on Writing Equations

Algebra 1 – *I'm Hair to Help* Unit   Name: _____

Writing Equations Practice   Date: _____ Class Period: _____

1. Jessie measured her hair on August 1 and again on September 1. Her hair had grown ¾ inch in that month. Assuming her hair grows at a constant rate and knowing that her hair is 13 inches long right now, what is an equation that could represent her hair length over time? If she were to not cut her hair for 6 months, how long would her hair be?

2. David just shaved his head. He has an average hair growth of 6 inches per year. Write an equation that could represent his hair length over time (in months). How long would it take David to grow his hair to 4 inches, assuming he does not cut it?

3. You and your family decide to rent a boat to have some fun on the lake while on vacation. There is a required deposit of $125 for the 8-passenger pontoon boat, plus it costs $109 per day. Write an equation to represent the cost of the boat over time. How much would it cost for you to rent the boat for 4 days? If all 8 family members were splitting the cost of the rental, how much would one person have to pay to rent the boat for 4 days?

4. Your electricity went out last night and you are afraid of the dark. You had a candle that was 5 inches tall and was burning at a rate of ¼ inch per hour. Write an equation to represent the height of the candle over time. How long did you have candle light during your blackout? Did it last you all night, or did you have to light another candle?

Sample Template 7.4: *I'm Hair to Help* Project Planning Form

Name of Project: I'm Hair to Help	
Designed by (Teacher Name[s and Email Address[es]): Julie Evans (jevans@bsd.k12.in.us), Diane Rodriguez (drodriguez@bsd.k12.in.us), Melodie Miller (mmiller@egreene.k12.in.us), Rita Cox (rcox@lssc.k12.in.us)	
Project Idea What is the issue, problem, or theme of this project?	Students are challenged to determine how long it would take them to grow their hair long enough to donate to Locks of Love, a nonprofit organization that provides hairpieces to financially disadvantaged children suffering from long-term medical hair loss. They must research the regulations on donating their hair to the organization, as well as average hair growth rates in order to determine how long it would take. Students present their findings and give a persuasive and informative speech that demonstrates why their findings are accurate and encourages others to donate their hair to the organization.
Topic(s) Addressed List one or more topics this project addresses.	Write linear equations, interpret the slope and intercepts of a linear model, draw a best fit line for a scatter plot and determine the slope of that line
Essential Question What is the Driving Question or challenge?	As philanthropists, how can we determine the amount of time necessary to grow our hair long enough to donate it to Locks of Love?
Entry Event What is the hook to launch this project?	Guest speaker: Someone who has lost hair due to a medical issue (and ideally has benefited from Locks of Love)
CCSSM and SMPs List those to be addressed by the project.	CCSSM: • HSN.Q.A.2: Define appropriate quantities for the purpose of descriptive modeling. • HSA.CED.A.2: Create equations in two or more variables to represent relationships between quantities; graph equations on coordinate axes with labels and scales. • HSF.IF.B.6: Calculate and interpret the average rate of change of a function (presented symbolically or as a table) over a specified interval. Estimate the rate of change from a graph. • HSF.IF.C.7.A: Graph linear functions and show intercepts. • HSF.BF.A.1: Write a function that describes a relationship between two quantities. • HSF.LE.A.1.B: Recognize situations in which one quantity changes at a constant rate per unit interval relative to another. • HSF.LE.B.5: Interpret the parameters in a linear or exponential function in terms of a context. • HSS.ID.B.6.C: Fit a linear function for a scatter plot that suggests a linear association. • HSS.ID.C.7: Interpret the slope (rate of change) and the intercept (constant term) of a linear model in the context of the data. SMP: • MP 1: Make sense of problems and persevere in solving them. • MP 3: Construct viable arguments and critique the reasoning of others. • MP 4: Model with mathematics.

Sample Template 7.4: Continued

T = Taught; P = Practiced; A = Assessed		T	P	A		**T = Taught; P = Practiced; A = Assessed**		T	P	A
Learner Outcomes Note the 21st-century skills taught, practiced, and/or assessed in this project.	Written communication	□	□	□		Technology literacy		□	■	□
	Oral communication	□	■	■		Work ethic		□	■	■
	Collaboration	□	■	■		Civic responsibility		□	□	□
	Critical thinking	□	■	□		Numeracy		□	■	□
	Information literacy	□	■	□		Core content skills		■	■	■

Habits of Mind Indicate one or two habits of mind that are the focus of this project	□ Persisting □ Managing impulsivity □ Listening to others □ Thinking flexibly	□ Thinking about thinking ■ Striving for accuracy and precision □ Questioning, posing problems □ Applying past knowledge	■ Communicating with clarity □ Gathering data, using all senses □ Creating, imagining, innovating □ Responding with awe	□ Taking responsible risks □ Finding humor □ Thinking interdependently □ Learning continuously

Presentation Audience

Student Production	Group products (major types):	Presentation to a group of peers, including a visual representation of their calculations (informative and persuasive)	**Check all that apply:** ■ Class ■ School ■ Community □ Experts □ Web (public) □ Parents ■ Other: Locks of Love representative, guest speaker
	Individual products (major types):	Peer and self-evaluations, individual calculations/graphs	

Sample Template 7.4: Continued

Assessments and Reflection	**Rubric(s)** Check and describe all that will be used for this project.	☐ Multimedia presentation rubric	■ Other: Math reasoning rubric
		■ Oral presentation rubric	■ Other: Work ethic and collaboration rubric
		☐ CCSS ELA and literacy writing rubrics	■ Other: Daily peer and self-evaluations
		☐ School writing rubric	☐ Other:
		☐ School learner outcomes rubric	☐ Other:
	Assessment Type(s) Check and describe all that will be used for this project.	☐ Quiz:	■ Performance assessment: Practice presentations and daily peer evaluations
		■ Test: At end of unit	☐ Notes review:
		☐ Essay:	☐ Checklist:
		☐ Online assessment:	☐ Concept map:
	Reflection tools Check and describe all that will be used for this project.	☐ Survey:	☐ Focus group:
		■ Discussion: Throughout project	☐ Personal learning plan:
		☐ Journal:	■ Student-teacher conference: Discuss individual calculations and graphs
Project Resources	On-site personnel:	Audience	
	Technology:	Internet access, projector, graphing software	
	Community resources:	Guest speaker, audience	
	Print resources:	Blank graphs, posterboard	
	Online resources:	Mathalicious website (www.mathalicious.com), Locks of Love website (www.locksoflove.org)	

Template adapted from the Buck Institute for Education (www.bie.org).

Sample Template 7.5: *I'm Hair to Help* Six Problem-Solving Phases Planning Worksheet

PBL Process Phase	NTKs	Scaffolding	Before Moving to Next Phase . . .
Phase 1 What is the need or problem?	• How fast does hair grow? • How are we forming groups? • What product is expected?	• Guest speaker • Driving Question • Research hair growth online • Homework: Review assessment rubric and bring questions to next class	1. Students can explain the problem during class discussion. 2. All questions from students have been answered during class discussion at the end of lesson 1. 3. Work ethic and collaboration rubric.
Phase 2 What needs to be in our solution?	• What is an average hair growth rate? • What do the slope and intercepts of the graph represent?	• Discussion and video of Ryan's hair growth over a year • Homework from lesson 1: • Answer questions about rubric • Continue research on hair growth	1. Students can find the average rate of hair growth. 2. Students can articulate what a quality product looks like according to the rubric. 3. Work ethic and collaboration rubric.
Phase 3 What are possible solutions?	• Can I use a graph? • Do I need to use an equation? • What representations do we need to include in our presentation?	Mini lesson on writing equations	1. Students can articulate their solution and explain how they arrived at the solution. 2. Work ethic and collaboration rubric.
Phase 4 Which solution should we use?	How do I generalize the solution?	Rubric criteria for math reasoning	1. Teacher's notes in observation checklists. 2. Work ethic and collaboration rubric.
Phase 5 How do we create, run, and inspect our solution?	• Is the information I am giving to my audience clear? • Can I prove that my solution is correct?	Use "I likes and I wonders" during practice presentations	1. Grade the practice run using the rubric. 2. Work ethic and collaboration rubric.
Phase 6 Reflect back: Did we solve the problem? Do we have a new problem?	• What questions might the audience ask us? • How do we field audience questions?	• Questions asked by the audience • Project rubric	1. Grade the presentation using the rubric. 2. Work ethic and collaboration rubric.

Template adapted from the Buck Institute for Education (www.bie.org).

Sample Template 7.6: *I'm Hair to Help* Lesson Plan for Day 1: Entry Event and Introducing the Project

Concept(s): *Possible medical background of guest speaker; empathy; basic discussion of how quickly hair grows; how to use a rubric*

Learning Targets: *Students should leave today with a better understanding of what the guest speaker has had to deal with (medically, emotionally, socially, and cosmetically). Students will work in pairs or small groups to devise a plan for calculating the average amount that hair grows (within a time period to be determined by each group of students).*

Standard/Indicators: *MP1*

Materials and Resources: *Guest speaker, rubric*

Three-Part Lesson Format

Note:
- Prior to the launch, discuss the guest speaker's background
- Have students think about and brainstorm appropriate questions for the guest speaker

Before / Launch (Getting Ready)
- Guest speaker (the "hook"—get students interested, and show the importance of the project)

During / Investigate (Students Work)
- Present the Driving Question: "As philanthropists, how can we determine the amount of time necessary to grow our hair long enough to donate it to Locks of Love?"
- Basic discussion of how to go about solving this type of situation (within groups)
- Listen carefully to their thoughts and ideas (and ask questions)
- Include questions to prompt students' thinking, such as, "What types of things do we need to know about our hair in order to know how long it will take to grow 10 inches?" and "What other important things do we need to know about our hair before we start to grow it out or before we cut it?"
- Observe and assess (on Days 1–3, I simply record a list of thoughts next to each student's name on a class list: What ideas and questions did this student have? What input did this student give? etc.)

After / Summarize (Class Discourse)
- Class discussion to bring ideas together
- Encourage a community of learners (encourage student-student dialogue, require justifications, call on students for their ideas, encourage students to ask questions, be certain that students understand what the teacher understands, move students to more conceptually based explanations)
- Accept students' solutions without evaluation
- Summarize main ideas and identify hypotheses

Homework
- Look over rubric and be ready to ask questions or make suggestions tomorrow

Sample Template 7.7: *I'm Hair to Help* Lesson Plan for Day 2: Estimating Hair Growth

Concept(s): *Estimation, graphing, slope, y-intercept, predictions, writing equations, line of best fit, measurement*
Learning Targets: *Students should be able to interpret the meaning of the slope and y-intercept. After watching the video, students should be able to apply their plan (from Day 1) for finding the average rate of hair-growth to predict how long it will take their hair to grow the required 10 inches for Locks of Love.*
Standard/Indicators: *F.LE.1.b, F.LE.5, F.IF.6, A.CED.2, N.Q.2, F.IF.7.a, S.ID.6.c, S.ID.7, MP1, MP3, MP4*
Materials and Resources: *Locks of Love video (from www.mathalicous.com) or search the Internet for a hair-growth time-lapse video, blank graphs, or graphing software (see Core Math Tools on the NCTM website [www.nctm.org/coremathtools/])*

Three-Part Lesson Format

Before / Launch (Getting Ready)
- Recall discussion from yesterday
- Rubric questions and suggestions

During / Investigate (Students Work)
- Work in groups
- Video (estimating hair growth)
- Graph hair growth from video
- Individual predictions and graphs
- Let go! (Let them struggle and let them solve the problem. Teachers should remind themselves that this is only Day 2, so students may not have a perfect understanding of what to do yet. It's OK; let them try to work through it together.)
- Listen carefully to their thoughts and ideas
- Include questions to prompt students' thinking, such as, "In our graph representing Ryan's hair growth, it seems like his hair continuously gets longer. What would our graph look like if he had cut his hair after two months?"
- Provide hints (with caution; avoid being the source of truth). Again, they will learn by struggling . . . let them!
- Observe and assess (on Days 1–3, I simply record a list of thoughts next to each student's name on a class list: What ideas/questions did this student have? What input did this student give to class and/or group discussions? etc.)

After / Summarize (Class Discourse)
- Class discussion to address questions

Homework
- Finish up any graphs or calculations
- Think about any issues that arose today (do students need a mini lesson to help them through those issues?)
- Think about presentation ideas

Sample Template 7.8: *I'm Hair to Help* Lesson Plan for Day 3: Putting It All Together and Preparing Presentations

Concept(s): *Estimation, graphing, slope, y-intercept, predictions, writing equations, line of best fit, measurement*

Learning Targets: *Students should be able to work as a team to design a multimedia presentation (using tools such as graphing software, PowerPoint, MS Word) that convinces the audience how long it will take for their hair to grow the required 10 inches for Locks of Love and strives to persuade them to do so.*

Standard/Indicators: *CCSS.ELA-Literacy.SL.9-10.4, CCSS.ELA-Literacy.SL9-10.5*

Materials and Resources: *Blank graphs, graphing software (see Core Math Tools on the NCTM website [www.nctm.org/coremathtools/])*

Three-Part Lesson Format

Before / Launch (Getting Ready)
- Pick up from where we left off yesterday
- Are there any issues? Do we need to have any mini lessons?

During / Investigate (Students Work)
- Work in groups
- Finish up individual predictions and graphs
- Prepare final product (graphs and presentation)
- Listen carefully to students final thoughts (and ask questions)—this will help prepare them for their presentation
- Include questions to prompt students' thinking, such as, "If I were an audience member who knew nothing about linear equations, would I understand your explanation of your graph?" and "Does your presentation clearly explain why it is so important to consider donating your hair to Locks of Love?"
- Provide hints (with caution—avoid being the source of truth)
    Note: Students should have a fairly solid understanding at this point. If they do not, more guidance may be necessary. Consider revisiting mini-lessons, if necessary.
- Observe and assess (on Days 1–3, I simply record a list of thoughts next to each student's name on a class list: What ideas/questions did this student have? What input did this student give? etc.)

After / Summarize (Class Discourse)
- Class discussion to address questions
- Discussion of presentation expectations

Homework
- Finish up presentations
- Be prepared for presentations tomorrow

Sample Template 7.9: *I'm Hair to Help* Lesson Plan for Day 4: Practice Presentations

Concept(s): *Estimation, graphing, slope, y-intercept, predictions, writing equations, line of best fit, public speaking, informative/persuasive speaking*

Learning Targets: *Students should be able to work as a team to design a multimedia presentation (using tools such as graphing software, PowerPoint, MS Word) that convinces the audience how long it will take for their hair to grow the required 10 inches for Locks of Love and strives to persuade them to do so. Each student should clearly understand his or her role in the presentation.*

Standard/Indicators: *CCSS.ELA-Literacy.SL.9-10.4, CCSS.ELA-Literacy.SL9-10.5*

Materials and Resources: *PowerPoint, posterboard, graphing software (see Core Math Tools on the NCTM website [www.nctm.org/coremathtools/]), any other materials for which students have indicated a need*

Three-Part Lesson Format

Before / Launch (Getting Ready)
- Final clean-up questions and discussion
- Reminder of what needs to be included in presentation (rubric)
- Reminder that individual predictions and graphs must be completed and turned in tomorrow

During / Investigate (Students Work)
- Observe students' practice presentations
- Score other groups, using rubric, and provide feedback using "I like" and "I wonder"
- Work in groups
- Final work time for groups to discuss feedback

After / Summarize (Class Discourse)
- Class discussion to address final questions
- One last reminder of presentation expectations

Homework
- Finish up presentations, using feedback given by teacher and other students
- Be prepared to present tomorrow
- Have individual products ready to turn in tomorrow

Sample Template 7.10: *I'm Hair to Help* Lesson Plan for Day 5: Presentations

Concept(s): *Estimation, graphing, slope, y-intercept, predictions, writing equations, line of best fit, public speaking, informative/persuasive speaking*

Learning Targets: *Students should be able to work as a team to engage the audience with a multimedia presentation (using tools such as graphing software, PowerPoint, MS Word) that convinces the audience how long it will take for their hair to grow the required 10 inches for Locks of Love and strives to persuade them to do so. Each student in the group should perform a role in the presentation.*

Standard/Indicators: *F.LE.1.b, F.LE.5, F.IF.6, A.CED.2, N.Q.2, F.IF.7.a, S.ID.6.c, S.ID.7, MP1, MP3, MP4, CCSS.ELA-Literacy. SL.9-10.4, CCSS.ELA-Literacy.SL9-10.5*

Materials and Resources: *PowerPoint, posterboard, graphing software(see Core Math Tools on the NCTM website [www. nctm.org/coremathtools/]) , any other materials for which students have indicated a need*

Three-Part Lesson Format

Before / Launch (Getting Ready)
- Final group discussions to get ready for presentations
- Introduce audience

During / Investigate (Students Work)
- Observe and assess students' presentations, using the rubric
  Note: The "Final Presentation" portion of the rubric is completed by both the teacher and the audience members. The rest of the rubric is completed by the teacher only.

After / Summarize (Class Discourse)
- Class discussion with audience
  Note: During this debriefing time, the audience gave individual groups positive feedback. This was a very positive discussion. Students left class that day feeling good about the project and what they had learned and shared with others.

Homework
- None

CHAPTER

8

To Netbook or Not to Netbook?

Austin Hauser, *Herron High School, Indianapolis, Indiana*

> *To Netbook or Not to Netbook* is designed to be a four-week project for an introductory-level statistics class. The unit focuses on a specific issue at the schools in this district: investigating whether the one-to-one netbook devices the school distributed are being used effectively. The students are tasked with designing a survey that will give them good data on the use of netbooks; they then distribute the survey, collect data, analyze the results, and present them to an authentic audience.

Standards Addressed

This project, which requires students to make inferences and justify conclusions from sample surveys, focuses on three Common Core content standards:

- Recognize the purposes of and differences among sample surveys, experiments, and observational studies, and explain how randomization relates to each (HSS.IC.B.3)
- Use data from a sample survey to estimate a population mean or proportion, and develop a margin of error through the use of simulation models for random sampling (HSS.IC.B.4)
- Evaluate reports based on data (HSS.IC.B.6)

While the project certainly contains elements of all eight SMPs, the most embedded practices are MP1: Make sense of problems and persevere in solving them, and MP3: Construct viable arguments and critique the reasoning of others.

Project Highlights

In the year before the project, Ben Davis University High School (BDU) in Indianapolis, Indiana, was chosen to pilot the district's proposed one-to-one device policy, and each student was given a netbook device. This project took place in the second year of the policy, so students and faculty were already familiar with the device. According to the principal, the district had considered many different options before deciding on a specific netbook, and the administration was interested in learning what the students and faculty thought about the device.

Relevance: Think about what problem or concern currently plagues the local community or the school. Then, if appropriate, use this problem situation to fuel students' learning of the content standards.

The project was done in collaboration with an eighth-grade math class from one of the district middle schools, whose students were studying basic statistics and had recently received netbooks. The high school students had to present the eighth graders with the task of using the survey they had designed to collect and analyze data from the middle school.

Spanning the district in this way required the high school students to learn some 21st-century skills that would not be present in a typical math curriculum: Not only did they have to solve the problem of getting the study to another school, they also had to use communication skills to ensure that the younger students understood how to conduct the study.

BDU is an early college high school, and the daily schedule is more like a college schedule. These students were relatively familiar with the PBL process, as many BDU teachers already employ PBL units within their curriculum.

The class met twice a week for four weeks. The 90-minute periods were advantageous for PBL, giving students ample time to work on the logistics of the project in class. However, having only eight class periods to complete the project meant that days often contained multiple workshops and activities, and buffer days in the schedule quickly filled up with necessary tasks. (See Sample Template 8.5: *To Netbook or Not to Netbook?* Project Calendar.)

Buffer days: Instructional days in a PBL unit where no instruction is planned. Make sure that the Project Calendar includes one or two free days, depending on the length of the unit, and use these days for students to catch up on work, put last touches on their products, or address any unanticipated NTKs.

Day 1: Entry Event

To launch the project, two guest speakers were invited: the school principal and the superintendent. The principal spoke to the students about what information the administration was hoping to glean from the study. The superintendent was not able to attend, but he sent a letter to the students, addressing them as "statistical researchers" and inviting them to present their findings to members of the Wayne Township administration (see Sample Template 8.1).

From there, the students brainstormed the "Knows" and "NTKs" needed to complete the project. Working in small groups of three or four, students participated in a workshop (a combination of vocabulary self-awareness and a jigsaw reading from the textbook) on some NTKs regarding surveys that the teacher anticipated they would ask, such as "How do we write good survey questions?" and "How do we survey properly?" (See Sample Template 8.3: *To Netbook or Not to Netbook?* Scaffolding NTKs.)

Jigsaw model: A teaching technique in which each small-group member is assigned a different subtask. Groups then "piece together" the overall task, with each member contributing her or his part.

There are many different ways that teachers could scaffold the NTKs for this project, but dividing the work across a small group made the overall task seem more manageable to students.

More details on this particular lesson and other related resources can be found on the NCTM website (www.nctm.org/more4u).

Rigor: For this unit, there was a need to introduce some vocabulary terms that appeared in the jigsaw reading (such as random sampling, convenience sampling, inference, and response bias) before moving forward with content (see Sample Template 8.7). This strategy not only kept students from getting hung up on a few words, but it also held them accountable for their own learning and gave them an organizational tool for keeping track of important terms.

Student Ownership

One of the most remarkable occurrences throughout the unit was the amount of ownership the students took in the project. Almost from the first day, they began adjusting the Project Calendar. The students often proposed more efficient ways to accomplish the tasks at hand. It was truly *their* project!

One significant adjustment was that students requested a change in the small-group structure; based on how the project developed, they asked to work on it as a class.

Relevance: Don't be afraid to let the students take control and direct the project. Giving them an opportunity to direct the project can greatly increase buy-in by letting them know this really is their project. This project ultimately differed greatly from the original plan, but all the changes were suggested and driven by the students themselves.

By the end of the project, the students had identified a problem, developed multiple solutions, come together to agree on a single solution, tested the solution, and created a very professional presentation for the administration. (See the example in fig. 8.1.)

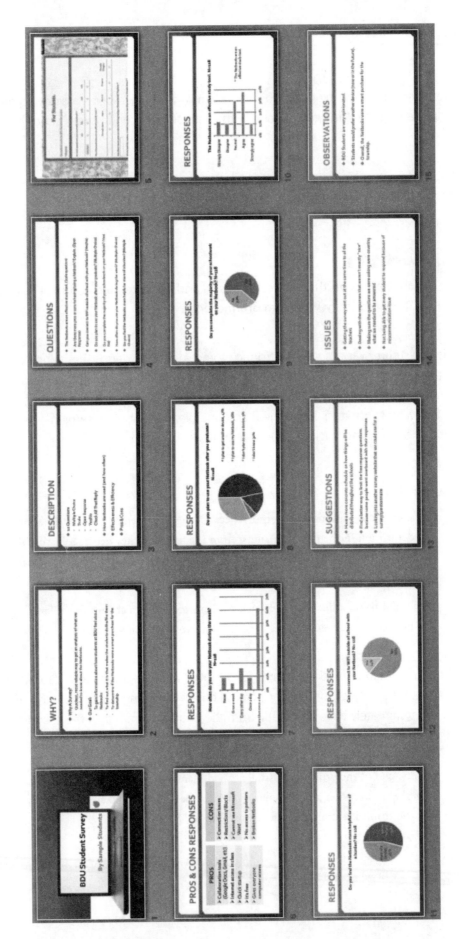

Fig. 8.1: Sample student work: Final presentation.

To conclude the project, the students participated in a Socratic-style discussion where they discussed successes, failures, and possible improvements to the project. In this reflection activity, the students were asked to focus on the statistical content of the study. At this point, they were the statisticians designing the project. Through this project, students were able to develop important 21st-century skills while taking control of their own learning.

Sample Template 8.1: *To Netbook or Not to Netbook?* Entry Event

Students at BDU,

As you are well aware, the Metropolitan School District of Wayne Township began an initiative to give students across the district one-to-one devices. This initiative started last year with the distribution of netbook devices to all students at Ben Davis University High School (BDU) and was extended this year to include students at Lynhurst 7th and 8th Grade Center. As a district, we now need to investigate how these devices are being used and how we can better use them.

Here is where you come in: I am asking you, as statistical researchers, to design a cross-district study to assess how the netbooks are being used. I would like data not only on how the students are using their devices but also on how teachers are implementing their use in the classroom. I would also like to know parents' opinions regarding the use of the devices.

After designing and distributing the study at BDU, I would like you to present the study to a group of students at Lynhurst who will conduct the same study at their school. After receiving all of the study results, I would like you to present your findings to me as well as other members of the administration in Wayne Township.

This is no easy task, but I am confident that you can accomplish it. I look forward to seeing what you come up with and, most importantly, the results of your study. My hope is that we can get some excellent data as well as ideas on how to better use these devices.

Sincerely,

Jeffrey K. Butts, Ph.D.
Superintendent of Schools

WE ARE WAYNE

Sample Template 8.2: *To Netbook or Not to Netbook?* Project Planning Form

Name of Project: *To Netbook or Not to Netbook*

Designed by (Teacher Name[s] and Email Address[es]): Austin Hauser, austinchauser@gmail.com

Project Idea What is the issue, problem, or theme of this project?	*Students will design, distribute, and analyze the results of a study to assess the productive use of netbooks by both faculty and students*
Topic(s) Addressed List one or more topics this project addresses.	*Designing studies, using data wisely.*
Essential Question What is the Driving Question or challenge?	*How can we design a study in order to investigate the productive utilization of netbooks by students and staff?*
Entry Event What is the hook to launch this project?	*Show a commercial advertising netbooks (for example, the YouTube video titled Chromebook: For Everyone, Everywhere). Ask the students what they use their netbooks for and how their teachers use them. Present a letter from the superintendent and the principal asking students to study the use of netbooks in their school. Ideally, have either or both of them attend the Entry Event, which reinforces the need for such a project and increases buy-in from the students.*

CCSSM and SMPs List those to be addressed by the project.	*CCSSM:* • *HSS.IC.B.3: Recognize the purposes of and differences among sample surveys, experiments, and observational studies; explain how randomization relates to each.* • *HSS.IC.B.4: Use data from a sample survey to estimate a population mean or proportion; develop a margin of error through the use of simulation models for random sampling.* • *HSS.IC.B.6: Evaluate reports based on data.* *SMP:* • *MP1: Make sense of problems and persevere in solving them.* • *MP2: Reason abstractly and quantitatively.* • *MP3: Construct viable arguments and critique the reasoning of others.* • *MP4: Model with mathematics.* • *MP5: Use appropriate tools strategically.* • *MP6: Attend to precision.* • *MP7: Look for and make use of structure* • *MP8: Look for and express regularity in repeated reasoning.*

Sample Template 8.2: *Continued*

T = Taught; P = Practiced; A = Assessed		T	P	A
Learner Outcomes Note the 21st century skills taught, practiced, and/or assessed in this project.	Written communication	☐	■	■
	Oral communication	☐	■	■
	Collaboration	☐	■	■
	Critical thinking	☐	■	■
	Information literacy	■	■	■

T = Taught; P = Practiced; A = Assessed		T	P	A
	Technology literacy	■	■	■
	Work ethic	☐	■	■
	Civic responsibility	☐	☐	☐
	Numeracy	■	■	■
	Core content skills	■	■	■

Habits of Mind Indicate one or two habits of mind that are the focus of this project.	☐ Persisting ☐ Managing impulsivity ☐ Listening to others ☐ Thinking flexibly	☐ Thinking about thinking ☐ Striving for accuracy and precision ■ Questioning, posing problems ☐ Applying past knowledge	☐ Communicating with clarity ■ Gathering data, using all senses ☐ Creating, imagining, innovating ☐ Responding with awe	☐ Taking responsible risks ☐ Finding humor ☐ Thinking interdependently ☐ Learning continuously

Presentation Audience

Student Production	Group products (major types):	*Study survey* *Survey results and analysis*	**Check all that apply:** ■ Class ■ School ■ Community ☐ Experts ☐ Web (public) ☐ Parents ☐ Other: Teacher
	Individual products (major types):	*Vocabulary worksheet* *Reflective journal entries throughout the project* *Summative quiz*	

Sample Template 8.2: *Continued*

Assessments and Reflection	Rubric(s) Check and describe all that will be used for this project.	■ Multimedia presentation rubric	☐ Other:
		■ Oral presentation rubric	☐ Other:
		☐ CCSS ELA and literacy writing rubrics	☐ Other:
		☐ School writing rubric	☐ Other:
		☐ School learner outcomes rubric	☐ Other:
	Assessment Type(s) Check and describe all that will be used for this project.	■ Quiz:	☐ Performance assessment:
		■ Test: *Individual post test*	☐ Notes review:
		☐ Essay:	☐ Checklist:
		☐ Online assessment:	☐ Concept map:
	Reflection Tools Check and describe all that will be used for this project.)	☐ Survey:	☐ Focus group:
		■ Discussion: *Socratic style*	☐ Personal learning plan:
		■ Journal:	☐ Student-teacher conference:

Project Resources	On-site personnel:	*Principal and superintendent as community partners.*
	Technology:	*One-to-one devices to create the survey and evaluate the results*
	Community resources:	
	Print resources:	
	Online resources:	

Template adapted from the Buck Institute for Education (www.bie.org).

Sample Template 8.3: *To Netbook or Not to Netbook? Scaffolding NTKs*

Scaffolding NTKs: Activity and Assessment Planning

Anticipated Knowledge and Skills Students Need (NTKs)	Assignment or Activity to Address NTKs	How Assignment or Activity Will Be Assessed	Learning Outcomes Addressed in Assignment/Activity
What type of study will we design?	Class discussion on day 1	Ask the students for their opinions.	Students will know we will be designing a survey-based study
What kinds of questions will we ask in the survey? How will we create these questions?	Group questions assignment	• Groups give a "mini-presentation" of their questions	Students will create a comprehensive survey to be sent out. Students will earn a grade based on their journal entries.
How will we analyze the results of the study?	Introduction to Excel/Google spreadsheet lesson (how to use functions, etc.)	• Formatively assess students during Spreadsheet lesson by walking around, ensuring they are on the right track when given short tasks • Exit slip at end of lesson asking students to give spreadsheet equations for certain statistical calculations (mean, median, standard deviation, etc.)	Students will be able to use ssurvey data
What information do we need to get out of the study?	Entry Event letter and introduction from community partner	Formatively assess through students' journals	Students will know what the administration wants to know about the usage of netbooks.
How do we appropriately sample? What are bad examples of sampling?	Lesson on sampling and surveying, using a jigsaw puzzle	Formatively assess through class discussion, using the jigsaw model	Students will know how to properly sample (and how not to).
How will we be graded throughout the project? How will peer and personal evaluations be considered in my grade?	• Rubric and assignment discussion on Day 1 • Project journal entries	Formatively assess through journals Note: If students show that they do not understand the evaluation process, it can be readdressed in later lessons.	Students will know exactly what is expected of them throughout the project.

Template adapted from the Buck Institute for Education (www.bie.org).

Sample Template 8.4: *To Netbook or Not to Netbook?* Six Problem-Solving Phases Planning Worksheet

PBL Process Phase	NTKs	Scaffolding	Before Moving to Next Phase . . .
Phase 1 What is the need or problem?	What information is the administration interested in obtaining through this study?	• Group contracts • Identify which NTKs are necessary to identify the need or problem	1. Students can explain the challenge or problem. 2. Students explain the problem to the facilitators.
Phase 2 What needs to be in our solution?	Whom do we need to survey? What information do we need to obtain from the survey? How do we create nonbiased and statistically appropriate questions?	Sampling vocabulary workshop	1. Groups explain and defend solution criteria (a student-generated checklist). 2. Groups compare and contrast their checklist with those of other groups.
Phase 3 What are possible solutions?	How do we survey everyone necessary? What questions should we ask?	Question compilation workshop	Students present multiple solution ideas based on criteria.
Phase 4 Which solution should we use?	What information can we appropriately glean from the data?	Rubric checks where students determine their score	1. Students articulate what solution they are going to do and defend it with the rubric. 2. Students consider resource constraints.
Phase 5 How do we create, run, and inspect our solution?	How do we present our results and suggestions to the administration?	Practice presentations with critical friends feedback via the rubric	1. How does the solution solve the problem? 2. Does your solution create new problem(s) to consider?
Phase 6 Reflect back: Did we solve the problem? Do we have a new problem?	How might we improve our survey? What questions did our results create? What questions did our results answer?	Socratic class discussion about the project	1. Expert/outsider compares final survey to project rubric. 2. Students reflect in their journal on what they learned.

Template adapted from the Buck Institute for Education (www.bie.org).

Sample Template 8.5: *To Netbook or Not to Netbook?* Project Calendar

Monday	Tuesday	Wednesday	Thursday	Friday
		WEEK 1		
	Day 1 • Entry Event • Discuss "Knows" and "NTKs" (shared in Google Docs) • PBL rubric and assessment discussion, including project journal description • Present basic format of the study (survey) • Divide class into small groups • Cover Sampling and Surveying using a jigsaw model • Define group roles • Assignment: Individually come up with at least 10 survey questions for discussion in the next class Goals: • Assess Knows and NTKs • Introduce sampling and surveys		Day 2 • Give some in-class time for groups to compile survey questions and decide on a final list of 10 in their groups • Compile 10 questions from each group • Discuss and narrow down survey questions • Begin compiling survey questions Goals: • Decide on the format of the survey (how many questions, what type of questions, etc.) • Compile all survey questions • Begin narrowing down and finalizing survey questions	
		WEEK 2		
	Day 3 • Finalize surveys • Decide on a plan to distribute the surveys • Conduct "Using Studies Wisely" workshop, with a discussion on how students might use their data • Decide how to split up the data evaluations • Treat remainder of the day as a buffer day Goals: • Finalize the surveys and come up with a distribution plan • Have a productive discussion on how to use studies wisely, with specific applications to this project		Day 4 Buffer day! • Begin analyzing survey data • Conduct workshop on using Google Spreadsheet to analyze results Goals: • Work on compiling and analyzing data and putting together a final presentation • Decide on presentation order for final presentations	

Sample Template 8.5: Continued

Monday	Tuesday	Wednesday	Thursday	Friday
		WEEK 3		
	Day 5 Buffer day Work on project presentation Goals: • Work on compiling and analyzing data and putting together a final presentation • Ensure that students know what material will be assessed on the upcoming test		Day 6 • Beginning of period: Practice delivering presentations to peers • Structured feedback from peers and reflection on presentation • Second half of period: Revise final presentations Goal: Students get productive feedback on their presentations and are able to make adjustments based on this feedback.	
		WEEK 4		
	Day 7 • Students present results to authentic audience • Results evaluated by authentic audience, teacher, peers, and self, based on the rubric • Discuss next steps: How did the project turn out? What can the authentic audience take from this project? Goals: • Show mastery of the study design and data analysis to the authentic audience • Reflect on the future implications of netbooks in schools		Day 8 • Summative assessment (quiz) • Personal reflection on project • Official peer evaluations • Conclusion discussion Goals: • Reflect on the project, gained knowledge, and future implications for the schools • Demonstrate mastery in the Designing Studies topics	

Sample Template 8.6: *To Netbook Or Not to Netbook?* Project Rubric

STUDENTS: _____
EVALUATOR: _____
DATE: _____

For each bullet point met, a box will be checked. Each checked box is equivalent to 1 point.

Criteria	Lacking Meets Some Requirements	Detailed Meets All Requirement	Thorough Surpasses Basic Requirements
Study Design (30%)	• Fewer than ten possible survey questions are submitted • Explains how at least one survey question will help answer the Driving Question • Contributes some to the survey question compilation conversation • Some effort to prevent bias is present in the survey questions ☐ ☐ ☐	• Ten possible survey questions submitted • Explains reason for using at least three survey questions — how it will help answer the Driving Question • Significantly contributes to the survey question compilation conversation, defending opinions with rational reasoning • Effort to prevent study bias is very clear and present in all survey questions ☐ ☐ ☐	• Eleven or more possible survey questions submitted • Explains reason for using at least five survey questions — how it will help answer the Driving Question ☐ ☐ ☐
Points (____ /10)	☐ ☐ ☐	☐ ☐ ☐	☐ ☐ ☐
Mathematical Content (40%)	• Makes some inferences from reports based on data • Spreadsheet activity completed • Data is mathematically presented in the presentation, with 70% accuracy • One mathematical function used (mean, median, mode, etc.) • Relevance of one mathematical function of the data mentioned (Why it's appropriate to use this function) • Contributes at least once to conclusion activity ☐ ☐ ☐	• Recognizes some of the purposes of and differences among sample surveys, experiments, and observational studies, and how randomization relates to each • Uses data from a sample survey to estimate a population mean or proportion • Evaluates reports based on data • Spreadsheet activity completed with at least 70% accuracy • Data in the presentation is at least 70% accurate • More than one mathematical function used. • Relevance of more than one mathematical function of the data mentioned • Contributes at least three times to conclusion activity ☐ ☐ ☐	• Recognizes the purposes of and differences among sample surveys, experiments, and observational studies, and how randomization relates to each • Uses data from a sample survey to estimate a population mean or proportion • Evaluates reports based on data • Spreadsheet activity is completed with 100% accuracy • Data are 100% accurate • Four or more mathematical functions are used • Relevance of all mathematical functions is mentioned • Contributes something to the Statistician's Viewpoint portion of the conclusion activity ☐ ☐ ☐
Points (____ /15)	☐ ☐ ☐	☐ ☐ ☐	☐ ☐ ☐
Work Ethic and Collaboration (15%)	• At least 70% of project journal entries completed fully • At least 70% of project journal entries are submitted by the due date • No negative comments from Peer Evaluations ☐ ☐ ☐	• Participates in all activities and workshops • All journal entries are completed fully • All journal entries are submitted by the due date • At least one positive comment from Peer Evaluations ☐ ☐ ☐	• Takes a leadership role in at least one activity or workshop • Positive comments from all group members on Peer Evaluations • Maintains a positive attitude throughout the project and is always on task. ☐ ☐ ☐
Points (____ /10)	☐ ☐ ☐	☐ ☐ ☐	☐ ☐ ☐

Sample Template 8.6: Continued

Final Presentation (10%)	• Data analysis is present • Seven of the following elements are present in the presentation: graphs, description of free-response results, sample survey, description of survey, why a survey is appropriate, statistical results, student vs. staff evaluation, observations, suggestions, and problems with the survey • Three group members verbally contribute to the presentation • Presentation is at least 5 minutes • Attempt to answer at least one question presented	• Data are presented professionally and without visible distraction • All 10 elements are present in the presentation • All group members verbally contribute; at least one group member contributes something not explicitly on the presentation materials. • Presentation is between 5 and 10 minutes • Accurately answers at least two questions from audience	• Multiple visual aids are present in presentation • All diagrams and pictures are clear and support the survey results • Presentation offers suggestions on how to improve the study • All group members contribute something not explicitly on the presentation materials • Accurately answers all questions from audience
Points (____/14)	☐ ☐ ☐	☐ ☐ ☐	☐ ☐ ☐

Feedback from the Judge

Please write down any remarks for the team's project in the appropriate section below.

Criteria	Comments
Design Content	
Mathematical Content	
Work Ethic and Collaboration	
Final Presentation	

Your final project grade will be computed as follows.

0.30*(____/10)   + 0.40*(____/15)   + 0.15*(____/10)   + 0.10*(____/14)   + 0.05*(____/20)   = ____/100
Design Content   Mathematical Content   Work Ethic and Collaboration   Final Presentation   Average of Peer Evaluations   Final Project Grade

Sample Template 8.7: *To Netbook or Not to Netbook?* Vocabulary Self-Awareness Sheet

Term	+, ✓, or –*	Definition	Example
Sample			
Population			
Sample survey			
Convenience sample			
Bias			
Random sampling			
Strata			
Stratified random sample			
Cluster			
Cluster sample			
Inference			
Sampling frame			
Undercoverage			
Response Bias			

* Key: + = I am confident that I can teach this term to someone else; ✓ = I know this term; – = I do not know this term

CHAPTER

9

"Worth"-Quake Insurance

Jeff Wilson, *Triton Central High School, Fairland, Indiana*

This four-week project was designed for an Algebra 2 course taught in a public high school that was being reconfigured as a school that exclusively used PBL as the instructional model. Students were thus familiar with group contracts, project rubrics, and other components of PBL before they began this unit. While most math courses at our high school were still taught in "traditional" style by single teachers for 50-minute periods, the math teachers were encouraged to design their own projects and implement them up to three times each semester to foster the school's PBL environment, develop deeper mathematical learning in all of our students, and help them persevere in solving much richer problems with true real-world applications.

Standards Addressed

"Worth"-Quake Insurance gives students a unique opportunity to use exponential growth graphs to make a compelling argument for purchasing earthquake insurance, and to make connections between exponential graphs and logarithmic graphs. As many as 15 CCSSM were addressed in this project, with strong emphasis on the following:

- HSF.IF.C.7.E: Graph exponential and logarithmic functions
- HSF.IF.C.9: Compare properties of two functions, each represented in different ways
- HSF.LE.B.5: Interpret the parameters in an exponential function in terms of a context

Through the display of graphs and corresponding data analysis in the limited space of a brochure, students are compelled to not only understand and create exponential-growth graphs but also to redesign them via technology so that they are visually appealing, mathematically viable, and understandable to a potential customer in an insurance office (see fig. 9.1).

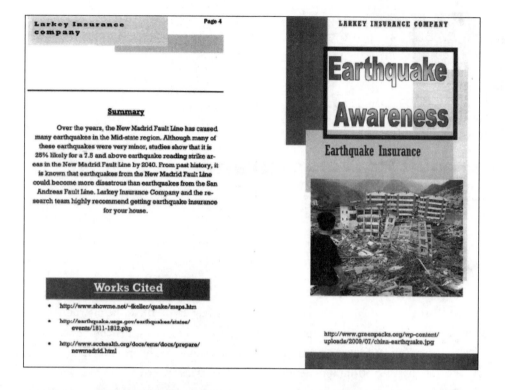

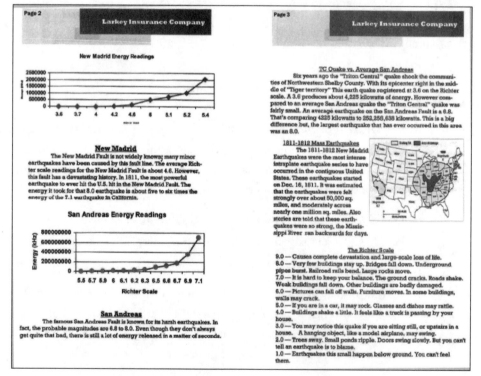

Figure 9.1: Sample student-produced brochure.

Students had to persevere through many requested design changes and graph reconfigurations for their brochures (MP1), reason quantitatively with precise results (MP2), and model their ideas and arguments with both graphs and appropriate written communication (MP4). (See Sample Template 9.4 for a detailed list of the CCSSM and SMPs addressed in this project.)

Project Highlights

The Driving Question of this project was, "As local insurance agents, how can we use current seismic data to convince clients to purchase earthquake insurance if the data indicate that an earthquake is imminent?" Students were engaged by the opportunity to create a brochure that could be used in a real-life setting.

The following overview highlights how the *"Worth-"Quake Insurance* unit is rigorous, is relevant to students, and helped students build relationships with one another.

Day 1: Entry Event

The project opened with a written request from a popular local insurance agent, our community partner, who also happened to be a highly successful graduate of our high school (see Sample Template 9.1). He needed a brochure that would compel the clients in his office to at least consider purchasing earthquake insurance, a hot topic in our community, as a couple of minor earthquakes had occurred in the past year along a previously unknown fault line.

Relevance: Hook students' interest in a long-term project by using current events they can relate to or are already concerned about.

All three of our Algebra 2 classes (approximately 90 students) embarked on this endeavor, in teams of three or four students.

Relationships: In the author's experience, nearly all projects function best with teams of three or four; teams of this size can evenly divide the work and manage one another's progress without having too much or too little to do.

Students were asked to include the following in their brochures:

- A scatterplot of earthquake activity on the famous San Andreas Fault in California over the past two decades, indicating Richter scale readings and the amount of energy released

- A similar scatterplot for the New Madrid Fault in the Midwest, including all earthquake activity in Indiana

- Some data analysis comparing the recent Triton Central earthquakes to recent California earthquakes

- Historical data about the strongest Richter measurements along the New Madrid Fault

Creating the Brochure: Rigor and Relevance

The request from the agent to include very specific earthquake data engaged students' critical thinking skills, as they sifted through websites and seismograph reports to determine what data were necessary to include for truly representational graphs of activity on both fault lines. Studying data along the New Madrid Fault created meaningful dialogue and a heightened sense of relevancy, as the students learned that the newly discovered fault line under our school district was part of the larger New Madrid Fault system, leaving most buildings in our district susceptible to severe damage if another earthquake matched the intensity of historical quakes along this fault line.

Many people in our community use this insurance company. Knowing that a "celebrity," the well-known insurance agent, would judge their presentations and their brochures—and select a couple of the best for use in his office—challenged students to produce a high-quality final product.

Working as a Team

Students developed and signed a group contract (see Sample Template 9.2) in which each member had to assess her or his own strengths and weaknesses. The contract delineated the different team roles (including San Andreas and New Madrid Surveyors, Teacher Liaison, and Brochure Designer); Teams had to agree on a work ethic and agree how to handle deadlines and manage conflicts. This type of self-management, in which students take responsibility for monitoring their own work and reinforcing the group's agreed-on expectations, is a key component of college and career readiness.

Relevance: Technology, writing skills, presentation skills, and research skills were all needed to varying degrees by each student on the team. Allowing team members to choose their role(s) really catered to students' various learning styles and strengths.

Rather than "just another worksheet," the contract was a vital component of the project and was in fact part of students' assessment. The Collaboration section of the project rubric (see Sample Template 9.3) required the group to discuss with the teacher how well the contract was maintained or broken. Group members assigned one another a point value for collaboration and work ethic, so group interaction and jigsaw learning were both vital to each group's success.

Scaffolding and Assessment

Most scaffolding occurred in the form of workshops or mini lectures and corresponding worksheets, roughly every other day of the project. This was sufficient for most students to learn the mathematics standards needed for the unit test (and state/national standardized testing) regarding exponential and logarithmic functions, and to understand the math needed to develop the graphs and the brochure. A few students required extra workshops while fellow group members worked on the brochure, which was easy to accommodate. NTKs about technology were actually handled by the students themselves—either among fellow group members or between groups. A few students even volunteered to lead brief workshops on technology issues as they arose.

Rigor: Having students lead brief workshops on technology allowed them to practice and hone their leadership skills and at the same time increased the complexity of the brochure designs.

Written communication was graded through nine journal entries (approximately one entry every two days—see fig. 9.2) and the final brochure.

1. What are your strengths and weaknesses as a student, in considering a role for this project?

2. What are the positives and the concerns that you perceive for your group at this point?

3. List three or more "rules" that an Algebra 2 student can use to accurately graph an exponential function, given its equation.

4. Graph on a coordinate plane your important Richter readings from your research (*x*-axis) in comparison to their corresponding earthquake energy levels (*y*-axis).

5. List the next steps for your group. Also list at least four more potential applications of exponential growth and decay, not including our earthquake project.

6. Make a rough graph representation of your Richter readings being placed on the *y*-axis and your corresponding earthquake energy levels on the x-axis. Which graph do you believe is better for the brochure: Richter readings on the *x*-axis or on the *y*-axis? Why? (Defend your group's choice for the graph with at least two solid mathematical reasons.)

7. What are the final steps necessary for your group to complete your brochure and prepare for your video of the brochure?

8. List at least three pieces of quality feedback your group received from critical friends today, which helped you readjust your brochure or video presentation. Who offered the feedback, and what adjustments did you make?

9. Write a paragraph addressing your group's ability to collaborate throughout the project. Include specific examples, and mention any issues you faced and how you resolved those issues. Write a second short paragraph that addresses your group's overall work ethic.

Figure 9.2: Journal entry prompts.

Oral communication was graded on the basis of the presentation of the brochure to the class and the insurance agent.

Week 2: Memo from Insurance Agent

The big "aha" moment of the project was introduced at about the halfway point. The insurance agent sent a memo to the students, saying he noticed that nearly every group had used the Richter scale for the *x*-axis and the corresponding energy levels released on the *y*-axis (an exponential growth graph). He expressed his concern that the layout of the brochure might require graphs that were more horizontal than vertical in nature, and he wanted to be prepared for both scenarios.

Students then had to redesign their graphs so that the *x*-axis and *y*-axis coordinates were reversed, which allowed us to talk about inverse functions and to introduce logarithmic functions. Students immediately understood what inverse functions were and how they have real-world applications, since they had already graphed the inverse functions of their exponential growth graphs. Thus, for the first time, the teacher had Algebra 2 students keen on learning the details of both inverse functions and logarithmic functions and their corresponding graphs.

Relevance: After hearing their children talk about this project at home and then seeing the students' final brochures, a few families purchased earthquake insurance for themselves!

Sample Template 9.1: *"Worth"-Quake Insurance* Entry Event

LARKEY INSURANCE
Todd Larkey, Fairland Agent

[Date]

Dear Triton Central Algebra 2 students,

No doubt, the extensive media coverage has made many of you aware of perceived increases in natural disasters. You are most likely also aware of the threats these disasters can pose to homeowners and local businesses, given the recent tornado touchdowns of the last four years here in Shelby County, as well as the devastating floods of June 2008. Local tragedies often prompt our home and business owners to seek more insurance coverage, which we strongly encourage before future weather threats occur.

One area of insurance coverage that most clients in Indiana fail to consider is earthquake insurance. Very few native Hoosiers have experienced the severity of a major earthquake or understand the Richter-scale measurements and the amount of energy unleashed by more potent earthquakes, such as those along the San Andreas Fault in California. Due to Indiana's lack of frequent earthquakes, home and business construction companies are not required to adhere to the same stringent building codes used in California. Therefore, most builders here in Indiana, in order to keep their construction costs more affordable for homeowners and businesses, choose not to earthquake-proof their buildings.

Indiana, however, does lie on an active fault line: the Wabash Valley Fault Line system, north of the New Madrid Fault along the Mississippi River. You may remember that Shelby County experienced two smaller earthquakes in the past few years, one of which had its epicenter in the Triton Central school district. While both earthquakes had minor Richter levels and did very little structural damage, the potential for more frequent and more severe tremors is possible and even probable. The U.S. Geological Survey has noted an increase in activity along the New Madrid Fault and has even sent a notice to many cities along this fault line to develop plans to deal with the consequences of an impending earthquake here in the Midwest.

We at Larkey Insurance want our clients to be ready for such a disaster. Our company would like a brochure to present to our clients that addresses their potential need of earthquake insurance coverage. The brochure should contain the following:

- Coordinate planes containing the earthquake activity on the San Andreas Fault over the past two decades, indicating Richter scale readings and energy released

- Coordinate planes containing the earthquake activity on the New Madrid Fault in the 20th and 21st century, indicating Richter scale readings and energy released

- Information that helps our clients understand the relationship between a Richter-scale measurement and the energy intensity of an earthquake

- Data analysis that compares the recent "Triton Central quake" to a "normal California quake"

- Data analysis that compares the recent Triton Central quake to a normal California quake

- Historical information about the strongest Richter readings ever experienced on the New Madrid Fault (1811–1812)

- A concluding section that encourages readers to carefully consider all the data presented (a well-written summary) and to seek advice on the costs and feasibility of earthquake coverage from his or her agent

We look forward to your group's presentation of your findings and brochure outline, which will occur on April 30, via a videoconference with our Fairland office staff. If a videoconference is not possible, your group may wish to submit a videotape/DVD advertising your brochure.

It is our hope to use the best brochure outline, or a combination of the best ideas from various groups, to produce a new earthquake coverage brochure for our central Indiana offices.

We thank you for your willingness to help us with this truly important civic matter.

Sincerely,

Todd Larkey

Sample Template 9.2: *"Worth"-Quake Insurance* Group Contract

Team Name_____   Start Date_____
                                                            End Date_____

Team Members

Name	Phone	Email	Times available after school

Member Assessments

Name	Strengths	Weaknesses	Strategies to improve weaknesses

Team Roles

Role	Name	Responsibilities
1. San Andreas Surveyor		
2. Teacher Liaison—the student representative for the group who usually has sole access to the teacher when project questions arise. The Liaison will also serve as an assistant to the San Andreas Fault Surveyor during research.		
3. New Madrid Surveyor		
4. Brochure Designer/New Madrid Assistant		

Personal Interactions
We agree to

1.

2.

3.

Work Ethic/Meeting Expectations
We agree to

1.

2.

3.

Managing Conflict and Absences

We agree to

1.

2.

3.

Violation of Agreement

If any part of this contract is violated, interventions may include the following:

- Teammate warning (please document in journal)
- Team intervention—group members assist student who is violating agreement (please document)
- Team meeting with teacher

If compliance is still not achieved, the violating student may be fired. All the individual's work will become the property of the group, and the fired student must complete all work individually.

Signatures

1. _____   2. _____

3. _____   4. _____

Sample Template 9.3: "Worth"-Quake Insurance Rubric

Criteria	Unsatisfactory (Below Performance Standards)	Proficient (Minimal Criteria)	Advanced (Demonstrates Exceptional Performance)
Content Knowledge (30)	• Shows evidence of some understanding of how to analyze exponential and logarithmic functions using different representations • Begins to develop proficiency comparing properties of two functions shows each represented in a different way (algebraically, graphically, numerically in tables, or by verbal descriptions) • Some evidence or interpreting the parameters in a linear or exponential function in terms of a context • Does not complete all required graphs in the brochure • Graphs had inaccuracies and/or incorrect best-fit lines • Written conclusions and/or data discussion do not accurately portray the data • Some supplemental worksheets are not completed, or those that are each contained several incorrect answers	• Shows evidence of understanding of how to analyze exponential and logarithmic functions using different representations. • Shows evidence of proficiency comparing properties of two functions, each represented in a different way (algebraically, graphically, numerically in tables, or by verbal descriptions). • Interprets the parameters in an exponential function in terms of the earthquake context. • Completes all required graphs • Graphs are accurate, with correct best-fit lines • Written conclusions and data discussion in the brochure accurately portrays the data findings of your group • All five supplemental worksheets are completed, each containing no more than a few incorrect answers	In addition to meeting the PROFICIENT criteria: • Shows evidence of deep understanding of how to analyze exponential and logarithmic functions using different representations. • Shows evidence of high level of proficiency comparing properties of two functions, each represented in a different way (algebraically, graphically, numerically in tables, or by verbal descriptions). • Interprets the parameters in a linear or exponential function in terms of a context. • Includes extra relevant information for the reader to consider (such as historical info on quakes, quotes from local homeowners or businesses who perused your info, extra mathematical concepts/graphs that prove your findings) • Graphs are well-defined and easily understandable for the reader, as based on your group's data findings • Written conclusions contain very persuasive arguments • Very few errors on any of the five supplemental worksheets
	0 - - - - 5 - - - - 10 - - - - 15 - - - - 20 - 22	23 - - - 24 - - - 25 - - -26 - - - 27	28 - - 29 - - - 30
Written Communication (20)	• Brochure contains portions that were poorly explained • Brochure contains several spelling and/or grammatical errors • Journal entries are incomplete or not well-detailed	• All portions of brochure are explained well and are easily understood • Brochure containes fewer than three spelling and/or grammatical errors • All journal entries are complete and well-written	In addition to meeting the PROFICIENT criteria: • Brochure is persuasive in arguments and in layout appeal • Brochure has no spelling or grammatical errors • Some journal entries contain extra insight, with more ideas discussed than were required
	0 - - 3 - - 6 - - 9 - - 12 - 14	15 - - 16 - - - 17 - - 18	19 - - 20
Oral Communication (20)	• Does not maintains eye contact • Use inappropriate body language, gestures, and/or facial expressions that distracted from delivery of the presentation • Inappropriate volume and enunciation for the presentation • One or two group members dominate the presentation • Not all members are appropriately dressed	• Maintain eye contact • Uses appropriate body language, gestures, and facial expressions for the presentation • Uses appropriate volume and enunciation for the presentation • All team members take part in the presentation • All team members are appropriately dressed	In addition to meeting the PROFICIENT criteria : • Body language, gestures, and facial expressions enhance the presentation • Volume and enunciation enhance the presentation • All team members are appropriately active in the presentation • All team members are professionally dressed
	0 - - 3 - - 6 - - 9 - - 12 - 14	15 - - 16 - - - 17 - - 18	19 - - 20

Sample Template 9.3: Continued

Criteria	Unsatisfactory (Below Performance Standards)	Proficient (Minimal Criteria)	Advanced (Demonstrates Exceptional Performance)
Collaboration (15)	• Fails to meet some of the group contract agreements • Fails to document critical feedback in journal entry • Fails to implement multiple items suggested in feedback from fellow group members and/or critical friends	• Meets all group contract agreements (documented in journal, plus teacher observation) • Documents critical feedback in journal entry • Implemented 2 or 3 items suggested in feedback from fellow group members and/or critical friends	In addition to meeting the PROFICIENT criteria: • Actively help other group members meet their group contract agreements (documented in journal, plus teacher observation) • Implements more than three items suggested in feedback from fellow group members and/or critical friends, as documented in journal entry
	0 - - - 2 - - - 4 - - - 6 - - - 8 - - - 10	11 - - - 12 - - - 13	14 - - - 15
Work Ethic (15)	• Attends 10 days or fewer during 15-day project • Is actively engaged in only some group research and discussion • Fails to document work ethic achievements/issues in journal entry	• Attends 11 to 13 days during 15-day project • Is actively engaged in most group research and discussion • Documents work ethic achievements in journal entry • Addresses some work ethic issues that were discussed in journal entry	• Attends 14 or 15 days during 15-day project • Is actively engaged in all portions of group research and discussion • Addresses all work ethic issues that were discussed in journal entry
	0 - - - 2 - - - 4 - - - 6 - - - 8 - - - 10	11 - - - 12 - - - 13	14 - - -15

Sample Template 9.4: "Worth"-Quake Insurance Project Planning Form

Name of Project: "Worth"-Quake

Designed by (Teacher Name[s] and Email Address[es]): Jeff Wilson, jwilson@nwshelbyschools.org

Project Idea What is the issue, problem, or theme of this project?	Create an earthquake insurance brochure for a local insurance company, which includes exponential growth graphs demonstrating the severe increase in damage done by earthquakes as the Richter measurement increases. Use an adjustment in the brochure layout as an opportunity to introduce and teach inverse functions and logarithmic functions.
Topic(s) Addressed List one or more topics this project addresses.	Exponential Functions and graphs, logarithmic functions and Graphs
Essential Question What is the Driving Question or challenge?	• For students: As local insurance agents, how can we use current seismic data to convince clients to purchase earthquake insurance if the data indicate that an earthquake is imminent? • For the Algebra 2 teacher: What is the mathematical connection between exponential and logarithmic functions and the real-world relevance of both?
Entry Event What is the hook to launch this project?	The opening document was a signed letter "mailed" in a sealed envelope to each group of students, requesting an earthquake insurance brochure that the insurance agent could use in his office to show prospective insurance clients. Students were assured that this was a legitimate offer—and in fact, the agent did use several of their brochures at the conclusion of the project. Relevance: A real, authentic project that serves a company or community provides motivation for students to produce a high-quality product.

Sample Template 9.4: *Continued*

CCSSM and SMPs List those to be addressed by the project.	CCSSM: • HSN.Q.A.2: Define appropriate quantities for the purpose of descriptive modeling. • HAS.CED.A.1: Create equations and inequalities in one variable and use them to solve problems. Include simple exponential functions. • HSF.IF.B.4: For a function that models a relationship between two quantities, interpret key features of graphs and tables in terms of the quantities, and sketch graphs showing key features given a verbal description of the relationship. • HSF.IF.C.7: Graph functions expressed symbolically and show key features of the graph, by hand in simple cases and using technology for more complicated cases. • HSF.IF.C.7E: Graph exponential and logarithmic functions, showing intercepts and end behavior. • HSF.IF.C.8.B: Use the properties of exponents to interpret expressions for exponential functions. • HSF.IF.C.9: Compare properties of two functions each represented in a different way (algebraically, graphically, numerically in tables, or by verbal descriptions). • HSF.BF.B.3: Identify the effect on the graph of replacing $f(x)$ by $f(x) + k$, $kf(x)$, and $f(x) + k$ for specific values of k. Experiment with cases and illustrate an explanation of the effects on the graph using technology. • HSF.BF.B.4: Find inverse functions. • HSF.LE.A.2: Construct linear and exponential functions, including arithmetic and geometric sequences, given a graph, a description of a relationship, or two input-output pairs. • HSF.LE.A.4: For exponential models, express as a logarithm the solution to $ab^{ct} = d$, where a, c, and d are numbers and the base b is 2, 10, or e; evaluate the logarithm using technology. • HSF.LE.B.5: Interpret the parameters in a linear, quadratic, or exponential function in terms of a context. • HSS.IC.A.1: Understand statistics as a process for making inferences about population parameters based on a random sample from that population. • HSS.IC.B.6: Evaluate reports based on data. SMP: • MP1: Make sense of problems and persevere in solving them. • MP2: Reason abstractly and quantitatively. • MP3: Construct viable arguments and critique the reasoning of others. • MP4: Model with mathematics. • MP5: Use appropriate tools strategically. • MP6: Attend to precision. • MP7: Look for and make use of structure. • MP8: Look for and express regularity in repeated reasoning.

Sample Template 9.4: Continued

T = Taught; P = Practiced; A = Assessed		T	P	A		**T = Taught; P = Practiced; A = Assessed**		T	P	A
Learner Outcomes List the 21st-century skills taught, practiced, and/or assessed in this project.	Written communication	□	■	■		Technology literacy		□	■	■
	Oral communication	□	■	■		Work ethic		□	■	■
	Collaboration	□	■	■		Civic responsibility		□	■	■
	Critical thinking	■	■	■		Numeracy		■	■	■
	Information literacy	■	■	■		Core content skills		■	■	■

Habits of Mind Indicate one or two habits of mind that are the focus of this project.	□ Persisting □ Managing impulsivity □ Listening to others □ Thinking flexibly	□ Thinking about thinking □ Striving for accuracy and precision □ Questioning, posing problems □ Applying past knowledge	■ Communicating with clarity □ Gathering data, using all senses ■ Creating, imagining, innovating □ Responding with awe	□ Taking responsible risks □ Finding humor □ Thinking interdependently □ Learning continuously

Presentation Audience

Student Production	Group products (major types):	• Final graphs, evaluated by peer groups first, then teacher • Earthquake insurance brochure, evaluated by teacher and insurance agent • Video presentation (1- to 2-minute commercial) advertising the brochure and its components	**Check all that apply:** ■ Class □ School □ Community ■ Experts □ Web (public) □ Parents □ Other:
	Individual products (major types):	• 9 journal entries demonstrating understanding of math/project components, and providing teacher with critical feedback • 7 worksheets of practice problems, covering all standards for exponential and logarithmic functions (some standards are covered in the brochure project, and some are not) • 2 quizzes and 1 exam on the math standards	

Sample Template 9.4: Continued

Assessments and Reflection	**Rubric(s)** Check and describe all that will be used for this project.	☐ Multimedia presentation rubric	■ Other: self-generated rubric (included here)
		■ Oral presentation rubric	☐ Other:
		☐ CCSS ELA and literacy writing rubrics	☐ Other:
		☐ School writing rubric	☐ Other:
		☐ School learner outcomes rubric	☐ Other:
	Assessment Type(s) Check and describe all that will be used for this project.	■ Quiz: *Two quizzes during middle of project*	☐ Performance assessment:
		■ Test: *One post-project comprehensive exam*	☐ Notes review:
		☐ Essay:	☐ Checklist:
		☐ Online assessment:	☐ Concept map:
	Reflection Tools Check and describe all that will be used for this project.	■ Survey: *Comprehensive post-project assessment*	■ Focus group: *Critical friends day to improve brochure*
		☐ Discussion: *Socratic style*	☐ Personal learning plan:
		■ Journal: *1 entries throughout the project*	■ Student-teacher conference: Meet with each group after project to discuss the entire four-week process and assign a grade

Project Resources	On-site personnel:	
	Technology:	
	Community resources:	*Local insurance agency*
	Print resources:	
	Online resources:	*U.S. Geological Survey websites:* • *Earthquake lists, maps, and statistics (https://earthquake.usgs.gov/earthquakes/browse/)* • *Search Earthquake Catalog (https://earthquake.usgs.gov/earthquakes/search/)*

Template adapted from the Buck Institute for Education (www.bie.org).

Sample Template 9.5: "Worth"-Quake Insurance Scaffolding NTKs

	Scaffolding NTKs: Activity and Assessment Planning		
Anticipated Knowledge and Skills Students Need (NTKs)	**Assignment or Activity to Address NTKs**	**How Assignment or Activity Will Be Assessed**	**Learning Outcomes Addressed in Assignment/ Activity**
Technology issues: • How do we design and print a brochure using a computer program? • How do we graph and print various functions on the computer using graphing technology? • How can we find reliable sources for data?	Student-led workshops as needed during project	Brochure	• Students will be able to use appropriate computer programs to graph exponential and logarithmic functions. • Students will be able to develop a brochure on the computer using appropriate technology.
• What are exponential growth and decay functions and graphs (F-IF.7e)? • What is appropriate "modeling with mathematics" (MP4)?	Class workshop, then extra workshops and practice problems on request	Worksheet, quiz, post-exam, and brochure	• Students will be able to graph exponential functions, including those with base e. • Students will be able to evaluate exponential functions with base e.
• What is a real-world application for exponential growth and decay? • What is appropriate "modeling with mathematics" (MP4)?	Class workshop conducted with information provided in a journal entry by students	Worksheet, post-exam, and brochure	Students will be able to use exponential functions (and logarithmic functions) to model and solve real-life applications.
What are logarithmic functions and graphs (F-IF.7e)?	Class workshop, then extra workshops and practice problems on request	Worksheet, quiz, and post-exam	• Students will be able to evaluate logarithmic functions with various bases. • Students will be able to graph logarithmic functions, including natural log functions.
How do we solve exponential and logarithmic equations (A-CED.1)?	Class workshop and extra practice problems on request	Worksheets and post-exam	Students will be able to solve exponential and logarithmic equations, and use them to model and solve real-life applications.
How do we interpret graphs (F-IF.4) and interpret parameters in terms of context (F-LE.5)?	Class workshops and journal entries	Worksheets, journal entries, brochure, and post-exam	Students will be able to convert data into appropriate graphs and then interpret the parameters of the graphs, both in written communication and by using mathematical language and symbols.
• How do we evaluate reports based on data and graphs (S-IC.6)? • How do we "construct viable arguments and critique the reasoning of others" (MP3)?	Class workshops, journal entries, and critical friends' assessments	Worksheets, journal entries, and brochure	Students will be able to draw correct conclusions about earthquake data and the degree of need for insurance.

Template adapted from the Buck Institute of Education (www.bie.org).

Sample Template 9.6: "Worth"-Quake Insurance Six Problem-Solving Phases Planning Worksheet

PBL Process Phase	NTKs	Scaffolding	Before Moving to Next Phase . . .
Phase 1 What is the need or problem?	• How do we generate this earthquake insurance brochure? • Which group members are responsible for each component of the project?	• Group contracts • Identify which NTKs are necessary to identify the need or problem	1. Each group member can identify various components of the project. 2. Each group member knows which members are responsible for the various project components.
Phase 2 What needs to be in our solution?	• What are the agent's requirements? • Where do we find the data for the graphs? • How do we generate the graphs? • What other math is needed to complete the project?	• Student-created criteria checklist • Math standards workshops • Communication workshops • Student-created checklist of required criteria	1. Each group member can identify the specific criteria required by the agent. 2. Appropriate group members have collected all necessary data for brochure graphs. 3. Appropriate group members have working knowledge of graphing technologies necessary to generate graphs.
Phase 3 What are possible solutions?	• What is a viable layout for the brochure? • What does a viable graph for this brochure look like?	• Critical friends' feedback within the group's members	1. Group members have sketched (by hand or with technology) multiple layouts for the brochure, with consideration for data presentation, visual appeal, conciseness, and other necessary qualities. 2. Appropriate group members have generated correct exponential/logarithmic function graphs for their earthquake information.
Phase 4 Which solution should we use?	Which brochure/graph designs meet the criteria, are the most concise, and are the most visually pleasing?	Critical friends' feedback with peer teams	1. Other groups have evaluated the group's layouts, using critical feedback in the format of discussion or a self-generated rubric. 2. The group's members have selected the "best" brochure and graph designs, and made the necessary adjustments, based on the critical friends' feedback.
Phase 5 How do we create, run, and inspect our solution?	• What issues did the critical friends see in Phase 4? • Should we address these concerns, and if so, how?	Critical friends' feedback	1. Another student group has evaluated the final draft of the brochure, using the project rubric. 2. After receiving the critical friends' feedback, the group has made final alterations and has submitted the brochure to the teacher.
Phase 6 Reflect back: Did we solve the problem? Do we have a new problem?	• What does the group believe are the strengths and weaknesses of the brochure? • How did the agent/teacher/critical friends' feedback align with the group's strength-weakness assessment? • "What alternative solutions could we now pursue to meet the requests made by the agent/teacher?	• Journal entries • Expert and teacher feedback on brochure • Expert and teacher feedback on video presentation of brochure • Post-project survey	1. Outside experts (teacher and insurance agent) have checked the project and provided feedback. 2. Students have submitted journal reflections about what they learned. 3. Group has completed a post-project survey.

Template adapted from the Buck Institute of Education (www.bie.org).

Sample Template 9.7: "Worth"-Quake Insurance Project Calendar

Monday	Tuesday	Wednesday	Thursday	Friday
WEEK 1				
• Present Entry Event document • Newspaper articles • Present group contract • Journal entry #1 • Present project rubric	• Two video clips on California and New Madrid Fault quakes • Group contract finished • Class shares NTKs • Journal entry #2	• Address NTKs • Time for computer research • Time for brochure design discussion (may need two days here)	• 15-minute workshop on exponential growth and decay • Worksheet #1 • Journal entry #3 • Journal entry #4	• 15-minute workshop on Euler number "e" • Worksheet #2 • Brochure development time with group
WEEK 2				
• 10-minute Q&A session for Worksheets #1 and 2, then a 20-minute quiz • Journal entry #5	• Jigsaw activity based on Journal entry #5 input: Present the best examples of exponential growth applications • Worksheet #3	• Memo from agent about the need for horizontally aligned graphs • Express other concerns about brochures • Journal entry #6	• 20-minute workshop on logarithmic functions and graphs, driven from the realization that they are inverses of exponential functions • Worksheet #4	• 10-minute Q&A session on Worksheets #3 and 4, followed by 20-minute quiz • Turn in Worksheets #1–4 and Journal entries #1–6 • Group planning time
WEEK 3				
• Work all period on brochure • Hand out Worksheet #5 (due Monday) • Journal entry #7	• "Critical friends": Meet with all students in class who had the same role as you; discuss issues, ideas, and solutions about brochure • Journal entry #8	• Make brochure adjustments after getting critical friends input	• 20-minute workshop on logarithmic properties • Worksheet #6	• 20-minute workshop on solving exponential and logarithmic equations • Worksheet #7 • Turn in proposed brochure and Journal entries #7 and 8
WEEK 4				
• 15-minute Q&A session for Worksheets #5–7 (hand in at end of period) • Hand out unit study guide • Journal entry #9	• Work on study guide • Work on 2-min video presentation/commercial of brochure • Meet with each group to discuss brochures	• Q&A session for study guide • Work on 2-min video commercial of brochure • Revise brochures if necessary	• Unit exam (handed back on Monday; students have Tuesday–Friday of next week to retake it before or after school if needed)	• Final brochure submission • Final video submission • Videos shown in class

CHAPTER

10

SEL
connection

Advanced Algebra — Trig Functions Precalc

Super Baugh I: Flacco vs. Kaepernick

Lori Burch, *Bloomfield Junior/Senior High School, Indiana*
Enrique Galindo, *Indiana University Bloomington*

This unit brought together two unlikely components: the 2013 Super Bowl and biorhythmic cycles. Students used their new knowledge of biorhythms (physical, intellectual, and emotional) to predict which quarterback—Joe Flacco or Colin Kaepernick—would be the better quarterback during the 2013 Super Bowl.

The belief in biorhythms originated in the nineteenth century. Biorhythms are said to describe our daily lives, which are significantly affected by rhythmic cycles. There are three main cycles—physical, emotional, and intellectual—lasting 23, 28, and 33 days, respectively (Hines, 1998). In theory, knowing where one is in each cycle can help predict what kind of day one is likely to have. If we know a person's birthday and the number of days they have lived (the biorhythmic cycles are set to zero at the moment of birth), we can determine where they are in each cycle for any given day.

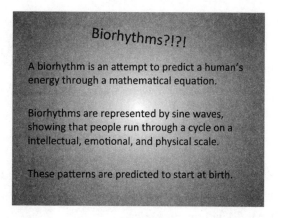

Fig. 10.1. A PowerPoint slide introducing biorhythms.

In 2013, for the first time in the history of the Super Bowl, the game featured two brothers coaching against each other: Jim Harbaugh of the San Francisco 49ers and John Harbaugh of the Baltimore Ravens. The game subsequently earned the nicknames "The HarBowl" and "The Super Baugh." The

2013 Super Bowl was memorable for other reasons as well, including the emergence of San Francisco 49ers quarterback Colin Kaepernick after quarterback Alex Smith was injured. This resulted in Kaepernick facing Baltimore quarterback Joe Flacco for the big game.

This unit engaged students in the study of trigonometric functions by having them create and interpret equations and graphs of biorhythms and use them to predict the performance of players in a high-stakes event.

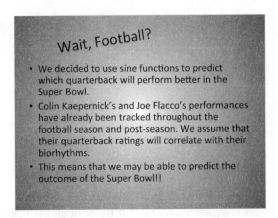

Fig. 10.2. A PowerPoint slide introducing the project.

The unit was implemented immediately after the two teams for the 2013 Super Bowl were in place, and designed to end in time for the class to watch the Super Bowl and see how their predictions played out.

Standards Addressed

This unit addressed three CCSSM:

- HSF.TF.B.5: Choose trigonometric functions to model periodic phenomena with specified amplitude, frequency, and midline.
- HSF.IF.B: Interpret functions that arise in applications in terms of the context.
- HSF.IF.C: Analyze functions using different representations.

The unit also addressed three SMPs:

- MP2: Reason abstractly and quantitatively
- MP3: Construct viable arguments and critique the reasoning of others
- MP4: Model with mathematics.

Project Highlights

The project's Driving Question was, "As sports analysts, how can we use biorhythms to predict who will be the better Super Bowl quarterback?"

Relevance: People love to pretend. Take advantage of this aspect of human nature, and allow students to take on a role other than a student in a mathematics classroom.

The Entry Event was a letter from a local columnist who commissions the class to provide a unique perspective for an article to be published before the Super Bowl.

This unit really opened up the kinds of conversations the teacher had with her students in the mathematics classroom. During the implementation of the unit, instead of talking strictly about mathematics, the class had wonderful conversations and debates centered around the NFL play-offs. After the Super Bowl, they were able to discuss how they saw the biorhythms playing out during the game. It was very exciting to see how students observed and analyzed each quarterback from the physical, intellectual, and emotional viewpoints.

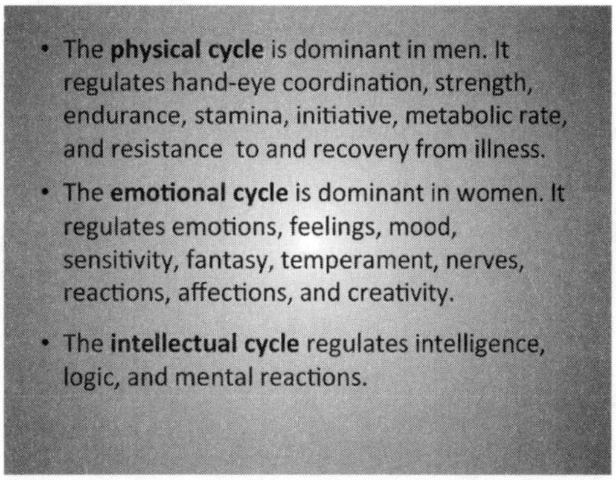

- The **physical cycle** is dominant in men. It regulates hand-eye coordination, strength, endurance, stamina, initiative, metabolic rate, and resistance to and recovery from illness.
- The **emotional cycle** is dominant in women. It regulates emotions, feelings, mood, sensitivity, fantasy, temperament, nerves, reactions, affections, and creativity.
- The **intellectual cycle** regulates intelligence, logic, and mental reactions.

Fig. 10.3. A PowerPoint slide from the project.

Relationships: PBL redefines the ways that teachers connect to students. Two years later, the teacher still has this connection with her students, all of whom agree that they have never watched the Super Bowl in the same way again.

Two things about how the teacher structured the presentations worked especially well:

- Instead of having multiple groups in the class rehash the same information over and over, each group had a different focus: either the physical, intellectual, or emotional biorhythms.

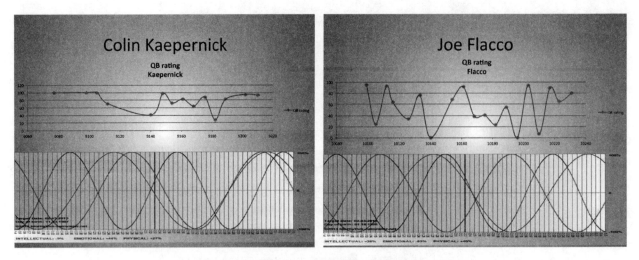

Fig. 10.4. PowerPoint slides from the project.

- • Students took turns presenting the material. For example, for a presentation on the emotional biorhythmic process, one student presented the information from Kaepernick's emotional biorhythmic curve, and another presented on Flacco's emotional biorhythmic curve. (See Sample Template 10.3: Instructions for the Physical Biorhythm Presentation.)

The guiding questions given to the other groups to help them prepare for their presentations are available on the NCTM website (www.nctm.org/more4u).

Rigor: Instead of having multiple groups present the same *mile* of information an *inch* deep each time, give each group its own focus for the presentation; this allows the class to cover the same amount of information in much greater depth.

Changes for Future Iterations of the Project

The teacher has a number of ideas about what she may change the next time she implements this project. Since this unit addressed only a few content standards and mathematical practices, she might either devote less time to it or include more standards and make the unit longer. This could be done by adding a statistics piece to the unit—analyzing the relationship between quarterback ratings throughout the year and the biorhythmic curves.

She also will not prioritize her own tastes and preferences when it comes to the aesthetics of presentation day. She still wants to ask community members, such as sports writers and coaches, to attend the class presentation, but she would make the event slightly less formal. The first time she presented the unit, her expectation was that students would dress professionally, and she had the final say on what "professional" meant. Next time, she will trust her students to take pride in *their* project, instead of imposing her own preferences on them.

Another interesting addition to this project would be to engage the students in deciding if there is any scientific evidence to justify biorhythms.

Rigor: When setting expectations for the project, think about what is really important. With details that are ultimately less important, challenge students to set expectations for themselves.

One suggestion for teachers interested in implementing this unit is to be aware of their audience. In the first days of implementing Super Baugh I, the teacher noticed that she was more excited about the project than her students were. Although she believes this is somewhat to be expected, as she had put so much time into preparing the unit, she asked the students why: "It's football! Aren't you excited?" One students responded, "You gave a football project to a bunch of nerds. Did you really think I watch football?" It was at that moment she remembered: not all teenagers have the same interests.

Relevance: Before designing a PBL unit, find out what the students are interested in. Listen closely to what they talk about during class, passing periods, or lunch, and make it a personal challenge to incorporate their interests. This will help make the PBL unit relevant to students and will engage them more fully.

College and Career Readiness Skill Building

The project not only supported important content in the Common Core, it also gave students opportunities to develop a set of learning habits and skills that are key to college and career readiness. For example, to succeed in the project, the students needed collaboration and communication skills. Working in groups, students with different ideas had to clearly communicate their thoughts to one another and then collaborate so that their individual ideas could come together in one unified group opinion. The group members then worked as a team to present this opinion to their audience.

In addition, throughout the process, groups and individual students had to regulate themselves. The crux of the project is predicting actual athletes' performances in the Super Bowl. It is not a project that can be extended if students need extra time. Expectations are stated, the timeframe is set at the beginning of the project, and students themselves must ensure that all expectations are met within the given amount of time.

Providing students with a list of questions that the teacher anticipates and would naturally ask may help groups edit themselves. Equipping students to question themselves in this way can aid their collaboration and self-regulation process for this project and beyond.

Rigor: Teachers should keep a list of questions they often ask their students and then challenge the students to make a list of questions as well. They may come up with things the teacher did not think of!

References

Hines, Terence M. "Comprehensive Review of Biorhythm Theory." *Psychological Reports 83* (1998): 19–64.

Sample Template 10.1: *Super Baugh I: Flacco vs. Kaepernick* Project Planning Form

Name of Project: *Super Baugh I*

Designed by [Teacher Name[s] and Email Address[es]]: *Lori Burch, lburch@bsd.k12.in.us*

Project Idea What is the issue, problem, or theme of this project?	*Students try to determine which quarterback of the final two teams will perform better at the Super Bowl. They research the history of biorhythms, and analyze the three main biorhythmic curves for each quarterback. Their final task is to present their research and analysis to a panel that includes sports enthusiasts (e.g., a columnist, television sports reporter, coach).*
Topic(s) Addressed List one or more topics this project addresses.	*Model the three primary biorhythms (physical, intellectual, and emotional) with sinusoidal functions. Graph each trigonometric function by hand and with technology.*
Essential Question What is the Driving Question or challenge?	*As sports analysts, how can we use biorhythms to predict who will be the better Super Bowl quarterback?*
Entry Event What is the hook to launch this project?	*A columnist commissions the class to provide a unique perspective for an article to be published before the Super Bowl.*
CCSSM and SMPs List those to be addressed by the project.	*CCSSM:* • *HSF.TF.B.5: Choose trigonometric functions to model periodic phenomena with specified amplitude, frequency, and midline.* • *HSF.IF.B: Interpret functions that arise in applications in terms of the context* • *HSF.IF.C.7E: Graph trigonometric functions, showing period, midline, and amplitude.* *SMP:* • *MP2: Reason abstractly and quantitatively.* • *MP3: Construct viable arguments and critique the reasoning of others.* • *MP4: Model with mathematics.*

Sample Template 10.1: Continued

Learner Outcomes	T = Taught; P = Practiced; A = Assessed				T = Taught; P = Practiced; A = Assessed		
		T	P	A		T	P	A
List the 21st-century skills taught, practiced, and/or assessed in this project.	Written communication	□	□	□	Technology literacy	□	■	□
	Oral communication	□	■	■	Work ethic	□	■	■
	Collaboration	□	■	■	Civic responsibility	□	□	□
	Critical thinking	□	■	□	Numeracy	□	■	□
	Information literacy	□	□	□	Core content skills	■	■	■

Habits of Mind				
Indicate one or two habits of mind that are the focus of this project.	□ Persisting □ Managing impulsivity □ Listening to others ■ Thinking flexibly	□ Thinking about thinking □ Striving for accuracy and precision □ Questioning, posing problems □ Applying past knowledge	■ Communicating with clarity □ Gathering data, using all senses □ Creating, imagining, innovating □ Responding with awe	□ Taking responsible risks □ Finding humor □ Thinking interdependently □ Learning continuously

Presentation Audience

Student Production		
	Group products (major types):	Presentation to a panel including the sports analyst commissioning the project. Presentation will include a visual representation of the calculations (informative and persuasive).
	Individual products (major types):	Peer and self-evaluations, individual calculations and graphs
		Check all that apply: ■ Class □ School ■ Community ■ Experts □ Web (public) □ Parents □ Other:

Sample Template 10.1: Continued

Assessments and Reflection	Rubric(s) Check and describe all that are used for this project.	☐ Multimedia presentation rubric	■ Other: Math reasoning rubric
		■ Oral presentation rubric	■ Other: Work ethic and collaboration rubric
		☐ CCSS ELA and literacy writing rubrics	■ Other: Daily peer and self-evaluations
		☐ School writing rubric	☐ Other:
		☐ School learner outcomes rubric	☐ Other:
	Assessment Type(s) Check and describe all that are used for this project.	■ Quiz:	■ Performance assessment: Practice presentations and daily peer evaluations
		☐ Test:	☐ Notes review:
		☐ Essay:	☐ Checklist:
		☐ Online assessment:	☐ Concept map:
	Reflection Tools Check and describe all that are used for this project.	☐ Survey:	☐ Focus group:
		■ Discussion: Throughout project	☐ Personal learning plan:
		☐ Journal:	■ Student-teacher conference: Discuss individual calculations and graphs

Project Resources	On-site personnel:	Audience
	Technology:	Internet access, projector, g raphing software (TI-Nspire calculators)
	Community resources:	Sports columnist, audience
	Print resources:	Blank graphs
	Online resources:	Google, Desmos (www.desmos.com)

Template adapted from the Buck Institute for Education (www.bie.org).

Sample Template 10.2: Super Baugh I: Flacco vs. Kaepernick Project Calendar

Monday	Tuesday	Wednesday	Thursday	Friday
		WEEK 1		
• Biorhythm Prezi (by Michael Crebbin) HW: Research Theory of Biorhythms.	• Know/Need to Know/Need to Learn • Coterminal angles HW: Begin to find NTKs (factual information)	• Sine equations and curves: Relationship between graphs and equation HW: Create three sine equations for both QBs (emotional, physical, intellectual)	• Sine equations and curves: using equations and graphs HW: Come up with two questions that could be answered using each graph/equation (i.e., two for emotional, two for physical, and two for intellectual)	• QUIZ: Sine equations and curves. • Do biorhythms seem valid for these 2 QBs? HW: DATA! Gather QB data to compare to biorhythms.
		WEEK 2		
• Graph sine equations. • Plot QB data over same time period. HW: Data analysis: • Do the data fit one of the curves better than any other? Which one? • How did you decide?	• Rank biorhythms • Assign groups for presentations • Delegate group responsibilities for group report HW: • Complete delegated portion of group report • Generate ideas for your group's portion of presentation	• Groups: Work on group report and group role in presentation. HW: • Revise delegated portion of group report. • Rehearse individual/group portion of group report.	• Practice presentation HW: Modify presentation based on suggestions from in-class practice.	• PRESENTATION: Note: Professional dress is required. If you do not own dress pants, let me know, and we'll figure something out. HW: Watch the Super Bowl. Bring snacks on Monday!

Sample Template 10.3: Instructions for the Physical Biorhythm Presentation

Team up with someone from the QB ratings group. Use their experience with Excel to generate two separate graphs of *your* biorhythm (Flacco and Kaepernick). (1 or 2 group members)

- "Insert Chart" —> "Scatter"
- *X*-axis = dates, *Y*-axis = biorhythm value (–1 to 1)
- Clearly title your graph (i.e., biorhythm, QB, . . .)
- Save your work. We will need it tomorrow.

Team up with someone from the other "physical" group. (1 or 2 group members)

- Do your curves (including the Super Bowl day) for each QB match?
- Use the following discussion questions to put together a 5-minute presentation. Each member of both groups must have a meaningful role.
  - What is the equation used to generate the physical biorhythm curve? Explain the "B."
  - How did we simplify the process to get around phase-shift issues?
  - Flacco:
    -- Display curve. Explain behavior from beginning of season to Super Bowl.
    -- What does it mean if he has a positive value? Negative value? Zero value?
    -- Does it make a difference if the value is trending upward or downward?
  - Kaepernick:
    -- Display curve. Explain behavior from beginning of season to Super Bowl.
    -- What does it mean if he has a positive value? Negative value? Zero value?
    -- Does it make a difference if the value is trending upward or downward?
  - Group prediction for better QB based on physical biorhythm. Is it a strong prediction or a weak prediction?
  - FYI: You may need to do some research to answer these questions. Your audience on Friday will not know about these biorhythms. It is your job to educate them.

Be thinking: We don't know which order the three biorhythm groups will present, but keep this in mind . . .

I think it would be cool to layer these curves, one on top of the other, as we proceed through the presentation. Ideas?

Sample Template 10.4: *Super Baugh I: Flacco vs. Kaepernick* Project Rubric

Super Bowl XLVII
SuperBaugh I: Flacco vs. Kaepernick

Student Name: _____
Group Members: _____

Criteria	Incomplete (Improvement Necessary)	1ST AND 10 (Proficient)	Touchdown (Exceptional Performance)
Group Process and Collaboration (20%)	• Does not adequately fulfill role within the group. • Is not actively engaged in all phases of the experiment. • Does not contribute (meaningfully, if at all) to group discussions. 0 - - - - - - - - - - - - - - 15	• Meets expectations for assigned role within the group. • Contributes meaningfully to group discussions. 15.1 - - - - - - - - - - - - - 18.4	In addition to meeting the 1ST AND 10 criteria: • Works with absent group members or helps struggling members within own group. • Facilitates direction for group work and discussions. 18.5 - - - - - - - - - - - - - 20
Work Ethic (10%)	• Misses 1 or more class sessions (unexcused) or does not make up for excused absences. • Does not attempt all outside homework assignments. • Is not actively engaged in all phases of the experiment. 0 - - - - - - - - - - - - - - 7.5	• Attends all class sessions, or makes up excused absences before the next class period. • Attempts all outside homework assignments. • Actively engages in all phases of the experiment. 7.6 - - - - - - - - - - - - - 9.2	In addition to meeting the 1ST AND 10 criteria: • Accurately completes (with written documentation) all outside homework assignments. 9.3 - - - - - - - - - - - - - 10
Presentation Aesthetics (30%)	• Eye contact not maintained. • Body language and facial expressions don't distract from presentation. • Volume is too loud or too soft for setting. • Words are not enunciated. • Pace of speech is difficult to understand. • Fails to demonstrate enthusiasm. • Shows little or no evidence of rehearsal. • Uses filler words: _____ • Relies on notes (reads from them). 0 - - - - - - - - - - - - - - 22.5	• Eye contact with audience is maintained. • Body language and facial expressions don't distract from presentation (maintains composure). • Volume is appropriate for setting. • Words are enunciated. • Pace of speech is understandable. • Demonstrates enthusiasm. • Shows evidence of rehearsal. • Avoids filler words. • Uses notes Sparingly (does not read from them). 22.6 - - - - - - - - - - - - - 27.6	In addition to meeting the 1ST AND 10 criteria: • Gestures, stance and movement enhance speech. • Presents with enthusiasm and clarity. • Polished delivery indicates rehearsal. • Does not read from notes. • Eye contact engages audience. 27.7 - - - - - - - - - - - - - 30
Mathematical Content (Presentation) (40%)	• Includes a scatter plot with no title or labels. • Does not explain why the data collected are periodic. • Is missing one or more components (amplitude, period, phase shift, vertical shift) of the curve. • Does not include an equation that accurately models the data set. • Does not explain the process used to find the best-fit equation. 0 - - - - - - - - - - - - - - 30	• Submits an accurate graph of the data collected for at least two periods, including title and labels. • Explains why the data collected are periodic. • Identifies components (amplitude, period, phase shift, vertical shift) of the curve. • Includes an equation that accurately models the data set. • Explains the process used to find the best-fit equation. 30.1 - - - - - - - - - - - - - 36.8	In addition to meeting the 1ST AND 10 criteria: • Graph includes at least five critical points. • Explains how each of the graph's components relates to the experimental data. • Explains the existence of multiple functions creating the same curve. 36.9 - - - - - - - - - - - - - 40

Sample Student Product on the Emotional Biorhythmic Process

THE EMOTIONAL BIORHYTHMIC PROCESS

By: Sample Students

WHAT IS A BIORHYTHM?

- ▸ A biorhythm is a cyclic pattern of physical, emotional, or mental activity said to occur in the life of a person.
- ▸ Modeled by curves on a graph.
- ▸ Shows the highs and lows of the everyday biorhythmic ratings.

Sample Student Product: *Continued*

THE EMOTIONAL CYCLE

▶ The curve that models a biorhythm usually has different sections based on the current day.

▶ On a negative emotional day, a person is supposedly more likely to react to situations in an unstable manner.

▶ While on a positive emotional day, a person is supposedly more likely to take situations in a more controlled manner.

THE EMOTIONAL SINE GRAPH

▶ The emotional cycle is modeled by the sine curve function $(sin(2\pi/28)x)$

▶ The Super Bowl quarterbacks Colin Kaepernick and Joe Flacco were born on different days, therefore they will have different emotional curves for the day of the Super Bowl.

▶ It is said that by calculating their emotional curves for that day, a better quarterback can be determined through the use of biorhythms.

Sample Student Product: *Continued*

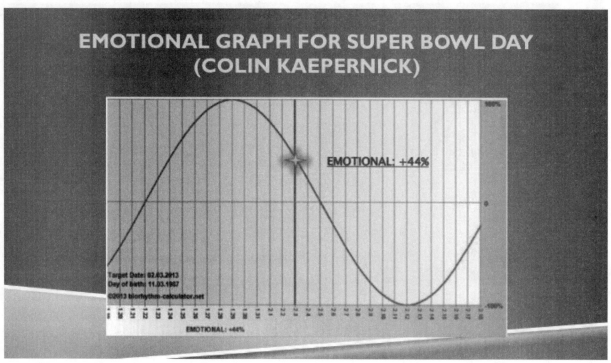

Sample Student Product: *Continued*

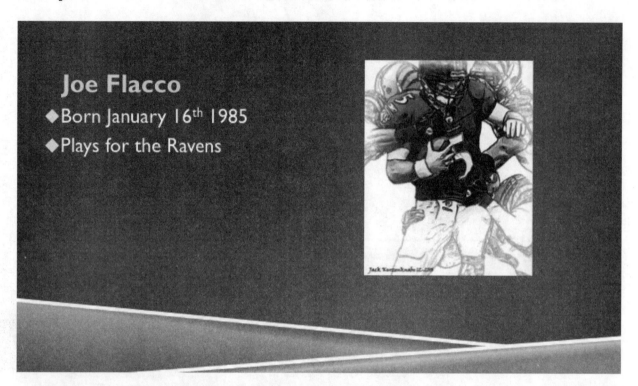

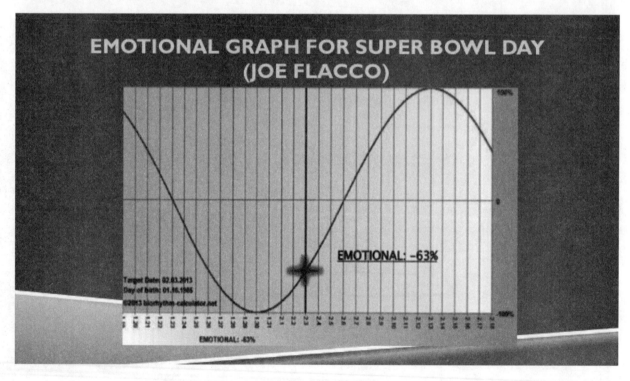

Sample Student Product: *Continued*

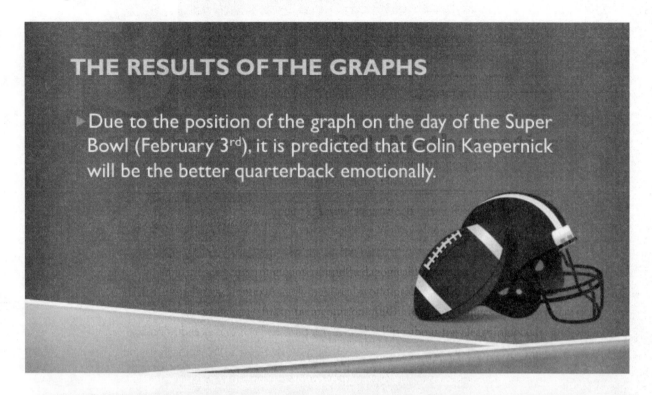

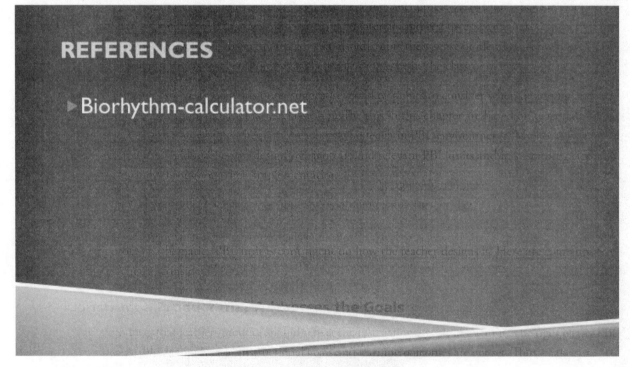

Sample Student Product: *Continued*

CHAPTER 11

Pre-Algebra/Algebra

figure out how to adapt this one ✗

Don't Sweat It!

Jacqulyn Fleming, *Indiana University Bloomington*

This project is a PBL unit in which students learn about data analysis and mathematical modeling through research on healthy eating habits and exercise routines.

Childhood obesity is a growing epidemic in the United States, and school is a primary resource for students to learn about what it means to be healthy. In fact, for many students, school is their only source of information about health and the only place they can eat a healthy meal and engage in regular physical activity. Government initiatives, such as Michelle Obama's "Let's Move" project, strive to provide healthy eating options and to limit the sale of junk food by placing restrictions on the kinds of food served in school cafeterias and at school functions. *Don't Sweat It!* gives math teachers and students the opportunity to work with school staff and administrators to raise awareness about these changing cafeteria guidelines. *might need to adjust this framing*

Standards Addressed

Students learn a number of relevant mathematical topics, such as using formulas, evaluating and interpreting expressions (HAS.SSE.A.1), and collecting, organizing, and interpreting data (HSS.ID.B5), all in the context of healthy living. The project also addresses four other CCSSM:

- Understand the concept of a function and use function notation (HSF.IF.A)
- Interpret functions that arise in applications in terms of the context (HSF.IF.B)
- Reason quantitatively and use units to solve problems (HSN.Q.A)
- Interpret linear models (HSS.ID.C)

Sample Template 11.1: *Don't Sweat It!* Project Planning Form shows the breakdown of the content standards and SMPs that are emphasized in this unit. Sample Template 11.2: *Don't Sweat It!* Scaffolding NTKs illustrates the scope and sequence in which these standards are uncovered, and how students' NTKs are embedded each week.

Project Highlights

The Driving Question for this project is, "Are certain types of food and exercise better for our health than others?" The teacher wanted her students to be able to explore the health-related topic of their choice, and she liked the open-ended nature of this question. Though the project was initially designed to address Algebra 1 standards and was intended for an older group of students, she modified it for her seventh-grade pre-algebra students, who adapted admirably to its heavy research components.

Day 1: Entry Event

The original Entry Event was to have the primary community partner (such as the school cafeteria manager, the principal, or a student counselor) pose the Driving Question and ask the students to help the school by creating informational displays to motivate their peers to make healthy food and exercise choices, to help the school transition to the changing cafeteria food guidelines, or to encourage qualifying students to sign up for the free and reduced-price lunch program.

[handwritten margin note: what other options could students look into! more choices for final product]

For her seventh graders, the teacher instead chose to launch the unit with the Chart Your Heart Rate Activity: Students were given data-collection sheets (see Sample Template 11.3) and tasked with monitoring their heart rate while engaging in a chosen physical activity. Working in the gym, students checked their heart rate every 5 minutes for a total of 30 minutes.

[handwritten margin note: incorporate mental/emotional health component]

Rigor: Attend to precision! Explain the details and procedures of the activity to the students in advance. Teach students how to find their heart rate, and have a plan for keeping track of time to get more accurate results.

Choosing an activity that was directly relevant to her students was a smart decision. Most of them were already very physically active in their extracurricular activities, and this activity gave them a useful life skill. Students later reported that the Entry Event was the highlight of the project.

Fostering Research Skills

Depending on how much time is available, the project can include up to three research phases before working on the final product:

- In Phase I, students investigate how exercise affects health.
- In Phase II, they research how diet (food intake) affects health.
- In Phase III, they explore how diet and exercise work together to affect health.

The teacher created a Project Research Collection Sheet to guide students in their investigations (see fig. 11.1).

Don't Sweat It! Project Research Collection Sheet

Directions: For this project, you will answer the following question: Are certain types of food and exercise better for our health than others? You will use the Internet to research your group topic and answer the specific questions you define.

Part I: Reword Your Topic and Find Resources

Reword Your group topic as a question
Example: Is running or swimming better for the health of your heart?
Record the online resources you find

Part II: Rewrite What You Learned About Your Topic

Write the answers to specific questions you researched
Directions: What did you learn in your research? Summarize the information in the space below.

Fig. 11.1: Snapshot of the worksheet.

Following the Entry Event, students learned how to enter their heart-rate data into an Excel spreadsheet and create a line graph. The students created individual graphs (see fig. 11.2 and fig. 11.3 for examples) and group graphs for comparison.

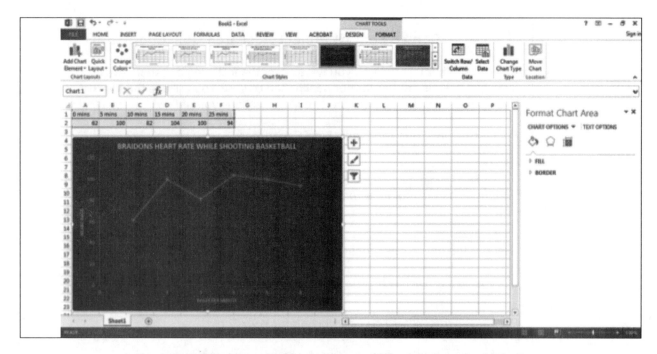

Fig. 11.2: Braidon's graph of his heart rate while shooting a basketball.

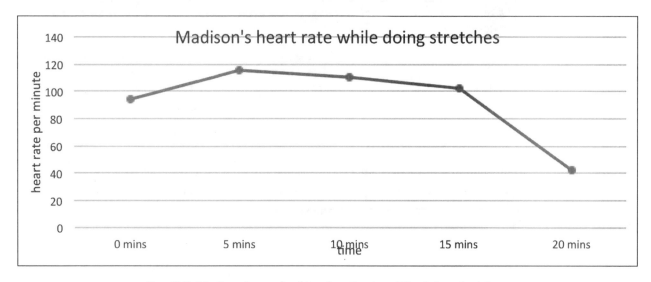

Fig. 11.3: Madison's graph of her heart rate while doing stretches.

Rigor: Post the students' line graphs around the classroom and ask students to make observations. Integrate the data-analysis standards for reading graphs into the daily classwork and homework.

Final Product

For their final product, students are asked to create an informational display, with the goal of answering the Driving Question in the context of a specific focus of their choice. For example,

- For a project focused on the new cafeteria guidelines, students could research the health-related reasons behind the guidelines and put together a display to raise the awareness of the student body.

- For a project focused on the free and reduced-price lunch program, students could encourage qualifying members of the student body to join the program by displaying information on the importance of a healthy diet.

For these seventh graders' final product, the students compared two or more foods or exercises and displayed the results of their research on a sheet of posterboard. (See fig. 11.4 and fig. 11.5 for examples.)

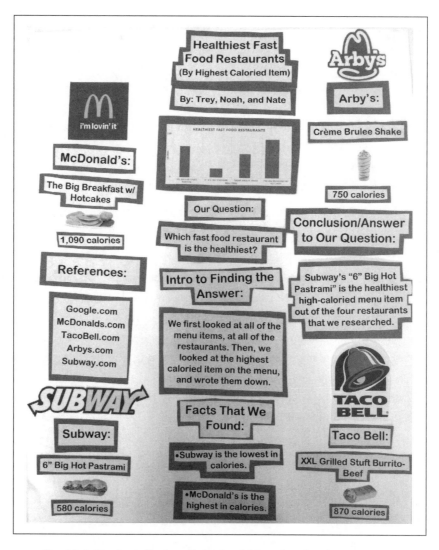

Fig. 11.4. Student display of the healthiest fast food restaurants.

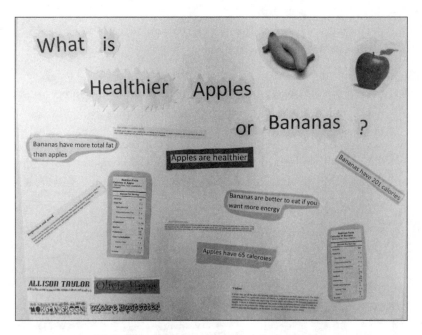

Fig. 11.5. Student display comparing apples and bananas.

Student researched such topics as "Do scary movies or video games raise your heart rate higher?" and "Does cheerleading or drill team provide a better workout?" They surveyed their classmates and used their knowledge of Excel to create charts based on the results of their research.

Possible Adaptations

This project can take on a variety of forms, from classroom-based to community-wide, and the Driving Question can seek either specific or open-ended results. For example, a more specific focus might be, "How can we raise awareness among students about the new cafeteria food guidelines?" or "How can we get more students who qualify for free or reduced-price lunch to sign up for it?" A more general focus might be, "How can we encourage healthy eating and exercise among students?" Teachers can also adapt the project to incorporate other subject-area standards, such as science, health, and physical education standards. Ideally, the project will be one that has rigor, is relevant to students' lives, and fosters relationships.

Sample Template 11.1: *Don't Sweat It!* Project Planning Form

Name of Project: *Don't Sweat It*	
Designed by (Teacher Name[s] and Email Address[es]): Jacqulyn Fleming, jacqulyn.c.fleming@gmail.com	
Project Idea What is the issue, problem, or theme of this project?	Students learn about data analysis and mathematical modeling through research on healthy eating habits and exercise routines. Specific variations on this broad topic include research into school cafeteria guidelines and the free and reduced-price lunch program. In groups, students create informational displays to motivate their peers to make healthy food and exercise choices.
Topic(s) Addressed List one or more topics this project addresses.	• Data collection, organization, and interpretation • Mathematical modeling • Formulas, expressions, equations, and graphs
Essential Question What is the Driving Question or challenge?	Are certain types of food and exercise better for our health than others? *Other options:* • How can we raise students' awareness of the changing cafeteria food guidelines? • How can we get more students who qualify for free and reduced-price lunch to sign up for it?
Entry Event What is the hook to launch this project?	Have the primary community partner (e.g., the school cafeteria manager, the principal, a school counselor) pose the Driving Question as a problem the school needs help with and ask the students to create informational displays to motivate their peers to make healthy food and exercise choices. • Depending on the specific Driving Question, school staff/administrators could ask the students to help the school transition to the changing cafeteria food guidelines or to encourage qualifying students to sign up for the free and reduced lunch program. • If there is no specific Driving Question, use the Chart Your Heart Rate Activity (see Sample Template 11.3).

Sample Template 11.1: Continued

CCSSM and SMPs List those to be addressed by the project.	CCSSM: • HSN.Q.A.1: Use units as a way to understand problems and to guide the solution of multistep problems; choose and interpret units consistently in formulas; choose and interpret the scale and the origin in graphs and data displays. • HSN.Q.A.2: Define appropriate quantities for the purpose of descriptive modeling. • HSA.SSE.A.1: Interpret expressions that represent a quantity in terms of its context. • HSF.IF.A.1: Understand that a function from one set (called the domain) to another set (called the range) assigns to each element of the domain exactly one element of the range. If f is a function and x is an element of its domain, then f(x) denotes the output of f corresponding to the input x. The graph of f is the graph of the equation y = f(x). • HSF.IF.A.2: Use function notation, evaluate functions for inputs in their domains, and interpret statements that use function notation in terms of a context. • HSF.IF.B.4: For a function that models a relationship between two quantities, interpret key features of graphs and tables in terms of the quantities, and sketch graphs showing key features given a verbal description of the relationship. • HSF.IF.B.5: Relate the domain of a function to its graph and, where applicable, to the quantitative relationship it describes. • HSS.ID.B.5: Summarize categorical data for two categories in two-way frequency tables. Interpret relative frequencies in the context of the data (including joint, marginal, and conditional relative frequencies). Recognize possible associations and trends in the data. • HSS.ID.C.7: Interpret the slope (rate of change) and the intercept (constant term) of a linear model in the context of the data. • HSS.ID.C.9: Distinguish between correlation and causation. SMP: • MP2: Reason abstractly and quantitatively. • MP3: Construct viable arguments and critique the reasoning of others. • MP4: Model with mathematics. • MP5: Use appropriate tools strategically. • MP6: Attend to precision.

Sample Template 11.1: Continued

Learner Outcomes
List the 21st-century skills taught, practiced, and/or assessed in this project.

T = Taught; P = Practiced; A = Assessed	T	P	A	T = Taught; P = Practiced; A = Assessed	T	P	A
Written communication	□	■	■	Technology literacy	■	■	■
Oral communication	□	■	□	Work ethic	□	■	□
Collaboration	□	■	■	Civic responsibility	■	□	□
Critical thinking	□	■	□	Numeracy	■	■	□
Information literacy	■	■	■	Core content skills	■	■	□

Habits of Mind
Indicate one or two habits of mind that are the focus of this project.

□ Persisting	□ Communicating with clarity
□ Managing impulsivity	■ Gathering data, using all senses
□ Listening to others	□ Creating, imagining, innovating
□ Thinking flexibly	□ Responding with awe
□ Thinking about thinking	□ Taking responsible risks
□ Striving for accuracy and precision	□ Finding humor
■ Questioning, posing problems	■ Thinking interdependently
□ Applying past knowledge	□ Learning continuously

Presentation Audience

Student Production

		Check all that apply:
Group products (major types):	In groups of three or four, students research a specific health- or exercise-related topic and create an informational display of their findings.	■ Class □ School ■ Community ■ Experts □ Web (public) □ Parents □ Other:
Individual products (major types)	Working individually, students maintain three journals: a heart rate journal, a food journal, and an exercise journal.	

Sample Template 11.1: Continued

Assessments and Reflection			
Rubric(s) Check and describe all that will be used for this project.	□ Multimedia presentation rubric	■ Other: *Informational display rubric*	
	■ Oral presentation rubric	□ Other:	
	□ CCSS ELA and literacy writing rubrics	□ Other:	
	□ School writing rubric	□ Other:	
	□ School learner outcomes rubric	□ Other:	
Assessment Type(s) Check and describe all that will be used for this project.	■ Quiz: *Students will take quizzes on math content.*	□ Performance assessment:	
	■ Test: *Students will take tests on math content.*	□ Notes review:	
	□ Essay:	■ Checklist: *Informational display checklist*	
	□ Online assessment:	□ Concept maps:	
Reflection Tools Check and describe all that will be used for this project.	□ Survey:	■ Focus group: *Students will provide feedback to their peers.*	
	□ Discussion:	■ Personal learning plan:	
	■ Journal: *Students will maintain three health journals*	■ Student-teacher conference: *The teacher will provide each group with feedback on their plan for the display*	

Sample Template 11.1: Continued

Project Resources	On-site personnel:	Students need access to the school gym or other recreational facility.
	Technology:	Students need access to the Internet, software (Microsoft Word and Excel), and a color printer.
	Community resources:	Students need access to school or community members knowledgeable about health and fitness.
	Print resources:	Students need access to books or other printed resources about health and fitness.
	Online resources:	Resources for Cafeteria Food Guidelines • Healthy Hunger-Free Kids Act (http://www.fns.usda.gov/school-meals/healthy-hunger-free-kids-act) • Nutrition Standards for School Meals (http://www.fns.usda.gov/school-meals/nutrition-standards-school-meals) • National School Lunch Program (http://www.fns.usda.gov/nslp/national-school-lunch-program-nslp, http://www.fns.usda.gov/sites/default/files/NSLPFactSheet.pdf) • School Breakfast Program (http://www.fns.usda.gov/sites/default/files/SBPfactsheet.pdf) • Let's Move—America's Move to Raise a Healthier Generation of Kids (http://www.letsmove.gov/learn-facts/epidemic-childhood-obesity, http://www.letsmove.gov/join-healthierus-schools-challenge) Activity Resources Note: This list is meant to be a starting point for each particular topic; it is not an exhaustive list of resources by any means. • Categories of exercise (http://go4life.nia.nih.gov/4-types-of-exercise) • General benefits of exercise (http://www.mayoclinic.org/healthy-living/fitness/in-depth/exercise/art-20048389, http://kidshealth.org/teen/your_body/take_care/exercise_wise.html#) • Maximum heart rate formulas (http://www.brianmac.co.uk/maxhr.htm) • Calories burned formulas (http://www.caloriesperhour.com/tutorial_BMR.php, http://nutritiondata.self.com/tools/calories-burned) • Explanations of what calories are (http://kidshealth.org/kid/nutrition/food/calorie.html, http://www.choosemyplate.gov/weight-management-calories/calories.html)

Template adapted from the Buck Institute for Education (www.bie.org).

Sample Template 11.2: *Don't Sweat It!* Scaffolding NTKs

Scaffolding NTKs: Activity and Assessment Planning

Anticipated Knowledge and Skills Students Need (NTKs)	Assignment or Activity to Address NTKs	How Assignment or Activity Will Be Assessed	Learning Outcomes Addressed in Assignment/Activity
How does exercise impact overall health? (Week 1) • What is heart rate? Why is it important? How is it calculated? • What are the "normal" ranges for heart rate, as related to age, weight, or other factors? • What are resting heart rate and maximum heart rate? • How fast does heart rate increase throughout the day? At night? During exercise? With stress (such as during a test)? With excitement? • How fast does heart rate decrease after exercise, stress, or excitement? • What does a healthy exercise plan look like?	Chart Your Heart Rate Students monitor their heart rate over time while performing a variety of exercises and activities, such as walking, running, and playing basketball. They create a chart or graph along with their peers who performed the same activity. As a class, students compare the charts of the various exercises and make observations and generalizations about the relationship between heart rate and exercise. The "Best" Equation for Heart Rate Students research formulas for predicting a person's resting heart rate and maximum heart rate. They will use the formulas for predicting their own heart rate and be encouraged to question how accurate the predictions might be. Since it is impractical to find each student's actual maximum heart rate, they can test the validity of the formulas from information about people whose maximum heart rate is known. Students will also look at intervals of "normal" heart rates and graph the information. Advanced students can define a piecewise function for the intervals of heart rates.	• Begin daily heart rate log (for example, record heart rate in the morning, afternoon, and at night). (May use phone apps). • Begin daily exercise log (example record any exercise done throughout the day). (May use phone apps). • Begin daily food log (example record all meals and snacks throughout the day). (May use phone apps). • Textbook problems on relations and functions, especially those displayed in chart form. • Textbook problems on domain and range of relations and functions. • Textbook problems on function notation, especially in the context of a formula (like the heart rate formulas). (This will also be a good time to review order of operations for simplifying an expression). • Textbook problems on graphing functions (e.g. graphing lines by using a table of ordered pairs).	• Students understand how data, personal or otherwise, are collected, organized, and interpreted. • Students understand the relevance of functions in mathematical models, such as the formulas for resting and maximum heart rate. • Students understand that mathematical models are not always exact, but rather are approximations of the world around us.

Sample Template 11.2: Continued

Anticipated Knowledge and Skills Students Need (NTKs)	Assignment or Activity to Address NTKs	How Assignment or Activity Will Be Assessed	Learning Outcomes Addressed in Assignment/ Activity
How does food intake impact overall health? (Week 2) • What are calories? How do we determine them? Why do they matter? • How many calories do I use (expenditure) in a day, a week, over time? • How many calories do I eat (intake) in a day, a week, over time? • How does increasing or decreasing my intake of calories impact my health? • What does a healthy diet look like?	*What Do I Need to Survive?* Students use formulas (found online) to predict their daily calorie expenditure (how much they are expected to burn off in a day), not taking into account exercise. They will take their daily food log (homework from previous week) and attribute calories to the foods they ate in order to predict their daily calories intake (how many calories they are eating in a day). They will then look at expenditure and intake together and asked essential questions about weight gain and loss. *What Are Calories?* Students research the calories in foods, creating charts to compare certain foods or food groups. They also research the meaning of a calorie, looking at the units involved and how calories relate to pounds. Extensions: (1) Ask students to alter a recipe they or their parents use at home to have less calories. (2) Invite the chemistry teacher to explain what a calorie is and do a science experiment of "burning" foods in order to determine the number of calories they contain.	• Daily food log revisited; students attribute calories to the foods they ate. • Continue to work on heart rate, exercise, and food logs of at least five days. • More textbook problems on function notation and plugging in variables to formulas. • More textbook problems on graphing functions by using a table of ordered pairs. • Textbook problems on solving formulas for a particular variable (and discussing the usefulness of this). • Textbook problems on unit conversions (e.g., converting from calories to pounds). • Textbook problems on solving linear equations in one variable. • Textbook problems on graphing linear equations in two variables by using slope-intercept form.	• Students understand how data, personal or otherwise, are collected, organized, and interpreted. • Students understand the relevance of functions in mathematical models, such as the formulas for caloric intake and expenditure. • Students understand that mathematical models are not always exact, but rather are approximations of the world around us. • Students understand the relevance of units in calculations and the necessity of being able to convert between analogous units.

Sample Template 11.2: Continued

Anticipated Knowledge and Skills Students Need (NTKs)	Assignment or Activity to Address NTKs	How Assignment or Activity Will Be Assessed	Learning Outcomes Addressed in Assignment/ Activity
How do diet and exercise work together to impact health? (Week 3) • How are exercise and calorie burn related? • How are heart rate and calorie burn related? • How do factors such as weight and age affect exercise and calorie burn? • How does lack of food affect our ability to exercise, stay awake, sleep, and function in life in general? • How does lack of food affect heart rate?	Balancing Fitness and Nutrition Students revisit their daily calorie expenditure to include exercise. They explore how including exercise at varying levels affects calorie intake and expenditure, and they explore the balance of food and exercise needed to maintain, lose, and gain weight, changing certain aspects as variables and keeping others constant. Extensions: (1) Have students design a personal health plan, incorporating diet and exercise. They can extend this plan to their view of a healthy lifestyle for a teenager and may also turn this into their final product. (2) Ask students what role heart rate plays in the exercise / diet equation. Are there other variables that should be considered when taking about nutrition and fitness? If so, how can we measure them and include them in our plan? Defining a Display Topic Once students have visited all of the previous activities and completed their logs, they can form groups and research potential display topics (keeping the driving question in mind). For example, students may choose to compare health facts of the old cafeteria foods with the new cafeteria foods. Once they have chosen a topic, students will write a proposal to the community partner addressing their idea in the context of the Driving Question.	• Complete heart rate, exercise, and food logs. (This must be completed prior to forming a group and developing a display topic proposal.) • More textbook problems on solving linear equations in one variable. • Textbook problems on problem solving with linear equations. • Problems modeling the amount of exercise needed to lose x number of calories. • In groups, create a list of potential display topics. • The final product idea must be written and reviewed by community partner prior to creation of display.	• Students understand how data, personal or otherwise, are collected, organized, and interpreted. • Students understand that mathematical models are not always exact, but rather are approximations of the world around us. • Students use their knowledge of functions to create a model for diet and exercise.

Template adapted from the Buck Institute for Education (www.bie.org).

Template adapted from the Buck Institute for Education (www.bie.org).

Sample Template 11.3: *Don't Sweat It!* Entry Event

Don't Sweat It! Project

Entry Event: Chart Your Heart Rate Data Collection Sheet

Name:_____ Class/Period: _____

Directions: Today, you will monitor your heart rate for a particular exercise or activity. First, you will determine your resting heart rate. You will then do your chosen exercise or activity for 30 minutes total and monitor your heart rate every 5 minutes. You will record your heart rate individually in the chart below, but you may work with others to accomplish the task. For example, you may play two-on-two basketball, stopping every 5 minutes to find your heart rate.

Part I: Choose an activity and make predictions.

Chosen Activity
Examples: Sitting, walking, running, playing basketball, dancing, stretching, playing jump rope, running up stairs

Predictions
Predict what will happen to your heart rate over time during your chosen activity. Be as specific as possible.

Sample Template 11.3: *Continued*

Part II: Record your heart rate and make observations.

Time	Heart Rate
0 minutes	
5 minutes	
10 minutes	
15 minutes	
20 minutes	
25 minutes	
30 minutes	

Observations

Make at least three observations. Consider the following questions:
- *What happened to your heart rate over time?*
- *How does your heart rate compare to those who did the same activity?*
- *How does your heart rate compare to those who did a different activity?*

Sample Template 11.4: *Don't Sweat It!* Six Problem-Solving Phases Planning WorkSheet

PBL Process Phase	NTKs	Scaffolding	Before Moving to Next Phase . . .
Phase 1 What is the need or problem?	• How will we research the latest restrictions on food in the cafeteria? • What percent of the student body qualifies for free and reduced price lunch and what percent of that population does not take advantage of the program? • How will we determine which group members are responsible for each component of the project?	• Identify which NTKs are necessary to identify the need or problem • Provide students resources to guide their initial research • Provide students a template for group contracts	• Students demonstrate their understanding of the Driving Question with a proposal for their display topic. • Students use the template to create a group contract.
Phase 2 What needs to be in our solution?	• What components need to be included in the informational display? • How will the display be graded?	• Generate a student-created checklist of required criteria • Provide students with a display rubric based on the criteria	Students use the provided checklist and rubric to guide their research and time management.
Phase 3 What are possible solutions?	• How will we know what information and sources on the Internet are accurate? • Once we gather sources, how will we organize it on the display?	• Offer workshops on valid sources from the Internet and on citing sources • Offer feedback on display ideas by way of the critical friends protocol (a structured procedure to provide feedback using the stems "I like, I wonder")	• Students sift through the research of the topic and begin to consider what to include in the display. • Students consider multiple layout options for their display.
Phase 4 What solution should we use?	How do we determine what information from our research to include on the display?	• Offer feedback on display plans by way of the critical friends protocol • Tell students to focus their feedback on what to include or exclud	• Students make changes to their display plans based on the critical friends feedback. • Students consider resource constraints.
Phase 5 How do we create, run and inspect our solution?	How will we know if our project answers the Driving Question?	Use the rubric to provide students with an expectation check	• Students make changes to their rough draft display based on the rubric check. • Students consider new solutions or more in-depth research questions not previously considered.

Template adapted from the Buck Institute for Education (www.bie.org).

Sample Template 11.5: *Don't Sweat It!* Project Rubric

Don't Sweat It! Project

Rubric and Assignment Details

Project Directions

For this project, you will answer the question, *Are certain types of food and exercise better for our health than others?* With your group, you will choose a particular topic to research in order to answer this question. Your group will display the results of your research on a piece of posterboard or a trifold display.

What to Include on Your Display

- A title
- Names of group members
- An introduction
- Visual aids and pictures
- Facts and information you discovered about your topic
- References for your facts and information
- Graphs (examples: pie chart, bar graph)
- A conclusion

Display Rubric

Your group display will be graded in the following categories.
- Content (60%)
- Written communication (20%)
- Creativity and innovation in design (20%)

(For specific details, see "What Does an Excellent Project Look Like?" below.)

Questions to Consider When Putting Together Your Display

- Does your title make sense based on your topic?
- Does your introduction tell your readers what your group researched?
- Are your visual aids and pictures relevant to your topic?
- Are your facts referenced? (In other words, your display should make it clear where the information came from.)
- Does your conclusion answer the question your group asked?
- Do the facts and information you provided support the conclusion you make about your topic?

What Does an Excellent Project Look Like?

Content Category (60%). The group—
- Fulfills all the requirements of the assignment
- Includes graphs, charts, or other visual aids created (with Excel or other software)
- Creates a professional and engaging display
- Displays accurate information that clearly shows the connection between the topic and the data incorporated

- Integrates the visual and graphical representations such that the information blends well with and heightens the message of the display
- Communicates the importance of the topic choice to future endeavors

Written Communication Category (20%). The group—
- Communicates with clarity and precision
- Cites claims (i.e., supports the information presented in the display)
- Adheres to the rules of spelling and grammar
- Shows awareness of the intended audience for the display

Creativity and Innovation in Design Category (20%). The group—
- Goes beyond the requirements of the assignment
- Creates a display that is visually appealing, well thought out, and interesting
- Demonstrates extraordinary evidence of original thinking, the creative process, and/or resourcefulness

Don't Sweat It! Project Display Rubric

Project Display Component	Possible Points	Points Earned
The group turned in a project.	20 points	
The names of the group members are included on the display.	1 point	
The display has a title.	1 point	
The title is relevant to the group's topic (i.e., the question the group is asking).	1 point	
The display has an introduction.	1 point	
The question asked or comparison made is clearly discernable from the display.	2 points	
The display includes visual aids (pictures) relevant to the topic.	3 points	
The display includes facts and information relevant to the topic.	6 points	
The display lists references (gives credit to others) for facts and information found.	1 point	
The display includes graphs (examples: pie charts or bar graphs) supporting the topic.	2 points	
The graphs were created by the group from data they collected (using Excel).	1 point	
The display has a conclusion.	1 point	
There is a discernible response to the question asked or comparison made in the display.	2 points	
The group pays attention to grammar and spelling (i.e., the writing is legible).	1 point	
The display is visually appealing, interesting, and engaging.	1 point	
Total	**44 points**	

12

Tips for Teachers from PBL Math Teachers

Sarah Leiker, *New Tech Network, Greendale, Indiana*
Jackie Fischvogt, *Greensburg Junior High, Columbus, Indiana*

As a PBL facilitator, you are about to become an expert in facilitation strategies, content standards, learning theory, career choices, and much more. Expertise, however, does not happen by creating one PBL project—or even two or three. As you refine your design and implementation processes, you will find that many of your initial failures ultimately evolve into improvements that lead to deeper learning experiences for students. At the same time, some pitfalls and obstacles are easy to anticipate, and taking some time to do so allows for a smoother transition into PBL.

In this chapter, we offer some tips on the design and implementation of PBL units, which will help you make the exciting jump into becoming a PBL facilitator.

Design

It's time! You've researched PBL, viewed sample mathematics PBL units, talked with PBL design "veterans," and dabbled in PBL creation—and now you're sitting down to craft your full PBL unit from start to finish. There are so many ideas and possibilities, and you're trying to sift through them all to know where and how to start your formal planning of a quality PBL unit—a unit in which students will engage in rigorous and relevant mathematical applications. You realize that not only must you carefully design the project context and scenario, you also need to plan how you will support students' understanding of standards and skills in a thoughtful and intentional way, so that students are prepared to explore those skills throughout the problem-solving process.

Below are some important issues to consider as you begin to craft a PBL unit and prepare for project implementation.

You Have So Many Ideas—Where Should You Start?!

You've been collecting ideas for project scenarios, thinking through the time you would like to devote to your project, and considering the standards and skills you want your students to become proficient in—now you need to find a way to sort through all of these ideas and start planning. Fortunately, working through the details of a project comes naturally to many mathematics teachers—we are problem solvers by nature!

It's often helpful to start with the Project Planning Form (see Sample Template 1.2: Project Planning Form in Chapter 1), which will remind you to "begin with the end in mind." This is a great way to brainstorm the content standards and 21st-century skills you want your students to uncover through their project research. It will also help you get started gathering potential project contexts that are authentic to the discipline, think about fostering connections to adults through real-world problems or challenges, and consider how to promote students' active exploration of the content to be learned.

 Once you've generated your big ideas, narrowing the focus of your Driving Question will become easier. The end result will be a possible project scenario that will fit more authentically in relation to the standards and skills you previously identified. Additionally, generating ideas for final products that require students to review data and information and to synthesize knowledge from their sources is a critical part of the brainstorming phase. Getting all your ideas in front of you will make the process of narrowing down the design details much easier as you continue planning.

Above all, remember that these exploration moments of the project design require you to be a learner of your content, of the authentic disciplinary connections of your content, of its real-world applications, and of assessment strategies. Remaining open to the possibilities available and to learning about those possibilities, instead of relying only on what you currently know, will ultimately result in a more robust project design.

If Student Voice Should Drive the Work and the Learning, How Much Day-to-Day Activity Needs to Be Planned?

In PBL, students are like explorers of an untraveled path. They are the ones learning to keep track of where they have been in a project, where they are now, and where they need to go next. They give and accept feedback in order to decide for themselves if they are headed in the right direction. Their interests are often the inspiration for a new project to explore, while their need for answers might lead to the creation of crucial, unanticipated workshops throughout a current project; and their insights allow for innovative approaches to a problem or unique solutions to a challenge. Student voice is essential to a successful project journey.

However, contrary to popular myth, PBL is not simply a go-with-the-flow teaching approach. Students are able to blaze new trails *because* their teacher has mapped out a solid course around a meaningful Driving Question and has included all the necessary components for students to be successful, such as benchmarks to check for understanding and progress. Without thoughtful, well-planned project components, students become lost, missing the chance to discover the heart of the project. Thus, day-to-day planning and mapping benchmarks are as important as remaining flexible and incorporating student voice.

While you begin to design the project components, know that you are creating the path for learning opportunities as students uncover the targeted content and skills necessary to respond to the project's Driving Question. It is critical that you establish this path *before* creating the actual Entry Event, rubric, summative assessments, and so on. Creating this path will ensure that the Entry Event introduces students to all the required problem-solving skills and that the rubric for assessment is all-encompassing.

As you develop your project plan and begin to map benchmarks, formative assessments, and scaffolding, consider whether your plan provides opportunities for students to do the following:

- Engage in inquiry and exploration of ideas
- Be creative
- Practice and develop a range of college and career readiness skills
- Make decisions about their learning (i.e., give students choices)
- Receive feedback and critiques from multiple sources
- Reflect on their learning and growth in a meaningful way
- Revise their work, working toward mastery

If you are planning to assess specific or targeted standards *and* skills, you need to design the ways in which you will teach both. Just as you wouldn't assess your students on their ability to solve systems of linear equations without teaching them methods for doing so, you certainly don't want to assess them on their ability to collaborate, for example, without teaching them how to contribute to a shared vision or make decisions together. Remember that students need scaffolds not only for content standards but also for skills such as organization and time management, using project calendars, meeting benchmarks, and even requesting workshops.

Planning the scaffolding for your project will prove to be one of the most important pieces of project design. (See Sample Template 1.1: Scaffolding NTKs: Activity and Assessment Planning in Chapter 1.) Thinking through the NTKs that students might generate and how those questions can drive students' learning of the standards and skills you've identified for the project investigation will ensure that you are supporting students' learning outcomes in every portion of the project. Spend time generating the anticipated NTKs for every targeted standard and skill, as well as the scaffolding activities students will engage in to find an answer to their NTKs, thereby learning the skills.

Is This Task Designed in a Way That Will Be Effective for Student Learning?

Your next step is to create the Entry Event and the rubric. Use the ideas from your pre-project brainstorms and the concrete project details you've planned to ensure that the Entry Event will elicit the NTKs you have anticipated from students.

When you have solid first drafts, review your Entry Event, rubric, and project plans to make sure the standards and skills you have targeted are sufficiently addressed. This will help to ensure that your own expectations for the project are in line with those stated in these documents, thus minimizing any surprises down the road.

The most important thing you can do during your planning process—and especially once you think your project design is ready to be implemented with students—is to get feedback on your work. Seek feedback from others in your professional learning network (e.g., colleagues, administrators, community partners, students, social media connections). You should also conduct a self-assessment, using the Six A's of Project Development (Markham, Larmer, and Ravitz 2003), which will result in a high level of rigor and authenticity in your project.

Authenticity. "Real world" is a phrase that many people like to use to describe authenticity. A true real-world project is one with an actual task or real-life problem, whose resources and adult contacts are all necessary to solve the problem.

- *Authentic task.* Mathematicians don't always *make products*; sometimes they find solutions, analyze information, or explore mathematical patterns. An authentic task may require students to apply mathematics to an exploration of real-world phenomenon (e.g., make predictions based on data, analyze quantities with the assistance of algorithms and theorems to investigate how things came to be, or predict what will happen in the future). Consider: In what ways do your project products and the individual assessments of student knowledge simulate the work of the mathematical discipline? In other words, do your project products resemble the work that mathematicians do? Mathematicians may not necessarily make a slide show or a pamphlet, but if students assume the role of a travel agent, they might. Look for an authentic product to fit the project's task and role.

- *Authentic sources.* What do mathematicians read and write about? What resources do they use to further their learning and critical thinking so they can apply their learning for new understandings? How can your students use these resources as well?

Academic rigor. To ensure that your project necessitates problem solving and conceptual understanding rather than "drill and kill" problem completion, review your project design with these questions in mind:

- Does your project require students to demonstrate learning derived from the CCSSM, your state standards, and/or your own learning outcomes?

- In what ways do your scaffolding plans address the anticipated NTKs and support students in developing content understanding and learning outcome skills?

- How is your scaffolding interactive and differentiated? How does it promote discourse? How do you use models?

Applied learning. A project that requires students to apply new skills and knowledge toward realistic, complex tasks engages them in finding a solution to the rigorous and authentic Driving Question of the project. This is a great moment to review your Driving Question and ask yourself, Are there several possible responses, solutions, and/or methods to this question? If so, you are one step closer to successfully designing a quality project in which students are required to apply their learning in meaningful ways.

Active exploration. Reflect on your plans for scaffolding and your current project calendar. Consider: How can students conduct research and inquiry into authentic sources in response to their NTKs? (Inquiry might involve having students create and experiment with mathematical models, or when appropriate, conduct their own independent research.) How can students have opportunities to make choices in regard to the actual direction of the project, not just, say, the formatting of the final product?

Adult connections. Don't underestimate the importance of outside adults to your students. Community partners can serve a number of purposes, big and small, throughout the project. For example, they may introduce the Entry Event, offer workshops, provide guidance on the process of creating the product, and evaluate the end products. Their presence lends authenticity, relevance, and a dash of excitement to the project, and depending on their level of involvement, can offer opportunities for students to make important connections with members of their community.

However, trying to integrate community partners into your project design can feel a bit overwhelming at first. Consider these tips as you begin the process of developing authentic community connections for your project:

- Start with friends, family members, neighbors, parents, or business persons you already have close connections with.
- Create a PBL "elevator pitch" to briefly explain what you're doing. Be prepared with suggestions as to how this person's expertise can affect the project work (e.g., as a guest speaker, to introduce the project and excite students about it, to serve on the panel that assesses students' final products). Consider your own NTKs regarding how to connect the person's real-world expertise to the rigorous standards of your PBL project.
- Most important: *Remember that this person, not you, is the expert in this field of work.* While you might be more of a PBL expert than the potential partners are, they are still the experts in their field. Take time to let them speak, and strive to be a learner of their profession—even if you have some expertise or experience in this area yourself.
- Take some time after your first meeting to think about the conversation before moving forward in the planning process. Then, rather than tell the community partners what the role will be, *collaborate* with them to determine how they might best work with you and your students.

Assessment practices. It is important to ensure that your project calendar contains multiple, varied, and authentic assessments related to the project throughout your unit. There should be opportunities for students to reflect on and revise their work and to receive feedback on their progress from teachers, mentors, and peers. Assessments should include an evaluation of the content standards and 21st-century skills reinforced throughout the project.

When reflecting on your rubric design, consider the following:

- How does your rubric assess the 21st-century skills and mathematical standards addressed in this unit?
- Does the final product allow students to demonstrate their ability to apply the knowledge and skills they have gained?

Implementation

You have designed your first project, full of authenticity, rigor, and exploration. You have made the time to request and receive design feedback from fellow teachers, to elicit exciting adult connections, and to create perfectly balanced assessments. Then the day arrives: Your first project is being initiated in a real classroom with real students who may be as novice to PBL as their teacher. Let us now address common difficulties experienced with first-time projects and some useful ways to avoid or correct these issues.

Students Did Not Ask the Right Questions after the Entry Event

You just launched your project with the Entry Event, and your your worst fear is realized: Thirty students remained silent and motionless, staring blankly.

Lack of engagement is a major problem for teachers, because the NTKs from the Entry Event guide all workshops, rubrics, assessments, and especially the project calendar. Here are some suggestions for engaging your students from the outset:

- Include multiple ways for students to engage in the launch. For example, your Entry Event could include both a video and a letter from someone in the community.

- Give students time to process their thoughts and ideas, both alone and with a group. Students may seem hesitant to buy in to the project, when they really just need some time.

- Use protocols to allow students to have a voice. Protocols such as chalk talks, think-pair-shares, and gallery walks maintain equitable communication and require students to share, listen, and think prior to giving feedback. Without such opportunities to voice their thoughts and opinions and to participate in a meaningful way, students disengage and wait to be instructed rather than take charge of their own learning.

- Encourage students to use highlighters or sticky notes to organize the new information into "Know" and "NTK" lists. Remember that students may have never used such lists before, and they will likely need your help with the process.

If these strategies don't work, ask your community partner for a follow-up conference call or email to encourage students, reinforce the authenticity of the project, and give them a gentle push in the right direction.

The Classroom Is Loud and Everyone Is Out of Their Seats

A PBL classroom becomes like a living lab and includes more movement and conversation than a traditional classroom. When everything is going right, the sounds of the room are more like a hum with an occasional heated debate. However, loud and disruptive behaviors impede learning in any environment.

Before the project begins, allow students to discuss barriers to their learning and how they learn best, and create a list of agreements and classroom norms together. Then enforce these agreements and norms, using group reflections and evaluations, group contracts, quick daily check-ins, easy-to-view posters, and other tools that meet students' needs. With each project, spend a few moments reflecting on students' new insights and ask if their classroom "pact" requires revisions or additions.

Eventually, student groups will create their own agreements and monitor the other members' contributions. The following is an example of how this might work:

A freshman student might acknowledge that he feels insecure about taking the lead and prefers to wait for a group member to assign him a task. This passive behavior leaves him stagnant much of the time. As a result, his group agrees that he should take the role of "liaison," which requires him to attend every group huddle with the teacher and the other group liaisons. After sharing on his group's progress in the huddle, he is then responsible for bringing feedback and next steps for revisions back to his group. He is no longer stagnant—he is in charge of a vital piece of the project.

The Students Either Don't Know What to Do, or They Are Taking over the Project, or They Don't Care and Aren't Taking the Project Seriously

As a first-time PBL designer and facilitator, you have given yourself permission to make mistakes; Give students that same allowance. When students don't understand expectations or how to resolve situations, they act just like adults and will take over, withdraw, become angry, or procrastinate.

While *you* may know where you're headed from Day 1, let the students join you in discovering their path. Listen to them, and let them guide their own learning experiences by using "Know" and "NTK" lists, daily next steps, and check-ins. This will help them process which scaffolding activities they need in order to learn the content and skills necessary to complete the project.

As you introduce these project-management tools, keep in mind that students do not inherently understand how to use scaffolds, such as rubrics, project calendars, and group contracts. Similarly, students likely don't have much experience with really listening to one another, managing time, communicating effectively, giving gentle reminders, or advocating for what they need. To help foster these skills among your students, take some time to reflect on the learning styles, collaboration, and facilitation strategies that support your own learning, and then model these strategies for your students. In addition, giving them immediate, positive feedback gives them the confidence to become better facilitators of their own learning.

Parents and Principals Don't See the Value in Our Work

Some teachers experience great support from their colleagues and regularly seek their feedback on project components, student work, and daily facilitation. Soliciting and using this feedback becomes an authentic way of improving project implementation and of modeling collaboration for students.

However, those who are not familiar with PBL may have questions and concerns about this nontraditional teaching method. Inviting people who are unsure about PBL to observe the classroom, evaluate student work, or act as community partners are all fantastic ways to help them find value in the students' projects.

Any time you ask for feedback or invite someone to visit your class, make sure to build in time to reflect on the feedback you are offered or to debrief the experience with the participants: How might you change your practice based on the feedback you have heard? How has the visit changed participants' views of PBL? These questions and others should engage stakeholders as part of an ongoing, open discussion.

Once Kids Finish the "Fun Parts," They Stop Digging Deeper

Students are likely to say "I'm done with my project" long before we think they are actually done. Truthfully, we are never "done" learning. Deeper learning requires higher-order thinking—it's not a matter of simply adding tasks to the original problem. Rather than have students do more work (such as having more advanced students help other students or do small tasks that add to their work), challenge them to take their learning and their work to a deeper level.

You might add a twist to the project that gives students a new problem to tackle—ideally, one that allows more opportunity to practice their new skills or provides an entry point into more advanced material. For example,

> In one PBL project, a group was given a generous budget to install carpet and tile and paint a room in a home. As students began to complete their calculations, they received an e-mail from the homeowner, who lowered the anticipated budget and added an additional task of replacing a semicircular window above the door. This allowed students to practice more of the same formulas they used for the flooring while exploring new formulas needed for the window. Because this twist was so realistic, the students didn't feel like it was extra work—instead, they dove right in to comparing pricing to fit the new budget, and applying the other skills they had learned to this new situation.

If students still feel "done" even after the added twist, give them time to evaluate their work using a rubric or to seek peer feedback on requirements they might have missed.

Last Words of Advice

PBL is an amazing adventure—but, as with any adventure, there are ups and downs. Some days will be overwhelming, haphazard, and strained. Other days, you will feel like you reinvented education and will wonder how you taught any other way.

To combat the down days, celebrate every success with students, recognize their (and your) accomplishments, and focus on why you started doing PBL in the first place.

> Our personal story of student success in PBL concerns a student whom we first met as an awkward seventh grader. While she qualified for advanced classes, her disorganization and lack of interest left her struggling to maintain B's and C's. After a year of PBL, her grades improved and she maintained solid test scores—but she also grew in many other ways. In groups, she began to passionately argue her viewpoint while learning to listen to others.
>
> Outside of school, a small-business project led her to start her own business with friends. After a few very long weekends of work, she had sold more than 250 bags of chocolate-covered pretzels, with total sales of approximately $1,000. As she began to create Facebook fan pages, she discovered a love for marketing; she carefully monitored her own data, hoping for more followers. If her numbers stalled, she designed competitions or trivia games to boost them. She soon hit her goal of reaching 2,000 followers from all over the world.
>
> Her newfound ability to articulate her opinions and present her thoughts led her to sit on student panels at summer institutes and to lead workshops for adults on how to train students to lead their own workshops.

The best part is that she began asking for the rubric on the first day of every project. She had learned to organize herself and to maintain her own learning.

PBL offers opportunities for students to master standards-based content in a real-life setting and to become navigators of their own learning. Whether for advanced students who are finally meeting their potential, or struggling students who use their hands-on expertise to win the rocket-launching competition, or the average shy students who are suddenly advocating for a community garden in front of a real audience, PBL creates a chance for students to discover who they are and where they want to be, both now and in the future.

Reference

Markham, Thom, John Larmer, and Jason Ravitz. *Project Based Learning Handbook: A Guide to Standards-Focused Project Based Learning.* 2nd ed. Novato, Calif.: Buck Institute for Education, 2003.

Tips for Teachers from a PBL Mathematics Educator

Jean Lee, *University of Indianapolis, Indiana*

PBL teachers must use their strong content and pedagogical knowledge in ways that support students' thinking in the context of solving a real-world problem. Transforming the learning environment to support successful implementation of PBL requires changes not only to curricula and instruction but also to teachers' and learners' thinking about how learning occurs and what mathematics entails.

It takes a long time to adopt new teaching practices and even longer to change habits of mind (Thornton, 2006). Shifting to a PBL approach can be a challenging endeavor for teachers, as it necessitates simultaneous changes in instructional materials and resources, pedagogical strategies, and assessment practices. For many teachers, it also disrupts their sense of identity—how they perceive themselves as teachers, and what their role should be in the classroom environment.

The PBL units showcased in this book suggest several key points to consider when designing and implementing PBL in a mathematics classroom. The tips in this chapter are based on personal experiences working with mathematics teachers who teach in PBL environments. My hope is that this chapter will aid teachers in designing rigorous and relevant PBL units and in sustaining rigor and relevancy throughout their implementation.

Design

A rigorous mathematics PBL unit is contingent on how the teacher designs it. Here are some tips for the teacher to consider.

Design or Construct a Product that Addresses the Goals

The products that learners produce should reflect the instructional goals of the unit. The rigor of any PBL unit can easily decline if clear instructional goals and learning outcomes are not set. Thus, make sure that the key mathematical ideas within the product guide the Driving Question and that the Driving Question is carefully crafted so that students learn the mathematics as they move through the unit.

Once teachers have set clear instructional goals and considered the big ideas that learners should demonstrate in the PBL unit, they should design the product they expect their learners to produce. This will help ensure that teachers have designed a rigorous PBL math unit that learners should be able to complete.

Supplement Textbook Problems with Good Tasks

The nature of the task plays a significant role in the kinds of mathematical engagement possible (Stein et al. 2000). Design or select cognitively demanding mathematical tasks—problems that promote conceptual understanding and the development of thinking, reasoning, and problem-solving skills (Stein et al. 2000)—that actively engage students in constructing and exploring mathematical concepts.

The term "cognitively demanding tasks" originates from Stein et al.'s Mathematical Task framework (Stein et al. 2000; Stein, Remillard, and Smith 2007), which describes a progression of how the actual written mathematical tasks are enacted through interaction with learners, teachers, and content. These tasks or activities are directly related to the context of the PBL and most importantly, should develop learners' competencies in mathematics. For example, if learners are creating and administering a survey for a statistics PBL unit, find rich activities and tasks that help students investigate the ideas of central tendency and sampling.

Schedule Assessments along the Way

Lesson planning should focus not only on the method of delivering mathematical concepts but also on selecting appropriate formative and summative methods of assessing learning outcomes (Colley 2008). These assessments should reflect the learning goals of the unit and appropriately measure to what extent learners are mastering the key mathematical concepts. This is particularly important if tasks or problems are immersed in the broader context of the PBL unit. Meaningful assessments contribute to maintaining a high level of academic rigor and active exploration.

Anticipate Misconceptions

Identify and anticipate possible learning challenges that learners will encounter within the unit, and likely student responses to cognitively demanding mathematical tasks (such as content misconceptions), in order to scaffold appropriately during implementation. Researching misconceptions that learners have regarding the mathematical concepts that teachers wish them to engage in can help sustain academic rigor and support learners in meaningful explorations (Smith et al. 2005), and allows teachers to include meaningful activities throughout the unit to better scaffold learners' understanding of these concepts.

This proactive researching has several additional benefits, including pointing teachers to rigorous tasks to incorporate into the project, building teachers' content knowledge, and increasing the productivity of whole-class discussions (Stein et al. 2008).

Implementation

A rigorous mathematics PBL unit is contingent on how the teacher implements it. Below is a list of tips for the teacher to consider.

Guide Learners to Make Mathematical Connections

It is important to support students as they engage in tasks that delve into the mathematics. However, many teachers find it difficult to support students' meaningful engagement with content without immediately giving them the answer. The challenge is knowing how to address students' claims and misconceptions in ways that do not lessen the cognitive demand of the tasks. Here are some ways to combat this challenge:

- Anticipate and identify learners' NTKs
- Focus on student thinking
- Support learners in demonstrating their understanding of the underlying mathematics with which they are engaging
- Ensure that learners are able to engage with accurate information

The teacher's ability to navigate among and balance between these practices enhances the academic rigor and active exploration of a math PBL unit.

Facilitate Productive Mathematical Discussion

PBL instruction reflects a classroom environment that is student-centered and inquiry-oriented. A student-centered mathematics class includes classroom discussions that become an important mechanism whereby learners are able to deconstruct or solidify their understanding of mathematical concepts. This kind of learning environment calls on teachers to effectively orchestrate productive discussions.

Hiebert et al. (1997) view the teacher's role in a discussion not as the main source of mathematical information nor the evaluator of correctness, but rather as a facilitator of conceptual understanding. The tasks teachers set are authentic mathematical problems that engage students in reflecting on and communicating about mathematics.

When students are working on selected tasks during a whole-class discussion, the teacher should facilitate, advise, guide, and monitor their thinking (Colley 2008). However, as noted by Stein et al. (2008), sustaining cognitively demanding tasks during whole-class discussions can be challenging. Several studies describe productive teacher moves that are important in sustaining the cognitive demand of tasks. Smith and Stein (2011) and Stein et al. (2008) and propose five teaching practices to make whole-class discussions more manageable:

- Anticipate students' responses to cognitively demanding mathematical tasks
- Monitor students' responses to the tasks during their exploration time
- Select particular students to present their work during the whole-class discussions
- Purposefully sequence the order in which students present their work

- Help the class make mathematical connections between different students' responses and key mathematical ideas

These five practices are designed such that each practice depends on the practice that precedes it. For example, how the teacher sequences students' responses in a whole-class discussion will depend on which students the teacher strategically selects to present their mathematical work.

Henningsen and Stein (1997) and Stein et al. (2008) found that keeping sustained pressure on students for explanation and meaning-making is an essential practice for maintaining high levels of cognitive demand in mathematics classrooms. "Keeping sustained pressure" means that the teacher holds students accountable for understanding a mathematical idea, regularly probing their thinking and pressing for reasons beyond quick answers. Facilitating discussions in this way sustains learners' opportunities to actively construct and explore mathematical ideas in a PBL environment.

Balance the Context and Content of the PBL Unit

One of the most important things to remember when rolling out a PBL unit is to maintain a healthy balance between the *context* of the unit and the mathematical *content* of the unit. Guiding learners to make connections among disjointed activities and important key mathematical ideas supports high-level mathematical thinking and reasoning (Henningsen and Stein 1997; Staples 2007). It is easy for learners to focus solely on completing the product in which the project in situated; remember to keep the learning of mathematical content in the foreground as well.

> For example, learners may learn about quadratics through a project that is situated in a projectile motion; learners also need to transfer and demonstrate their understanding of quadratics in other contexts that do not involve projectile motion.

Think about how the anticipated NTKs could be used as leverage to encourage learners to learn mathematical concepts. Stein et al. (2008) noted that a productive teacher practice in orchestrating whole-class discussion is for the teacher to help the class make mathematical connections between different class activities. Engle (2006) calls this "contextual scaffolding within PBL instruction" and says that teachers should support students in transferring the knowledge they gain in one setting to the project's context. Making connections between the content and the context of the project is extremely crucial in mathematics PBL instruction.

Reflect on the PBL Unit

After implementing a PBL unit for the first time, teachers should take time to reflect on how it went. Reflection is a process that requires a teacher to observe, analyze, interpret, and make decisions. According to Schön (1987), the reflective practitioner inquires into the nature of teaching, thinks critically about the work of teaching, and develops an understanding of when, why, and how to use particular approaches, techniques, and materials.

Teachers should reflect on what the students learned and what they themselves learned, as well as the progress learners made during the unit and the progress they made as a teacher. It may be easier to criticize what went wrong and note the improvements that should be made to a unit, but it is also important to analyze and document the progress the teacher and learners made throughout the project. Remember, PBL is an extended *process* of inquiry—take time to reflect on the journey!

Table 13.1 is adapted from the *PBL Starter Kit* (Larmer, Ross, and Mergendollar 2009) and can help start the process of reflecting on a PBL unit. The template uses the Six A's as a framework and walks the teacher through the strengths and challenges of the unit and future plans for improvement.

Table 13.1. Template for PBL reflection.

Six A's	What You Liked	What You Wonder	What You Plan as Next Steps
Authenticity • Complexity of the unit • Effectiveness of Entry Event • Authenticity/meaning to the students • Appropriate audience for students' work			
Academic Rigor • Quality and use of the Driving Question • Selection of the content standards • Scope and depth of central concepts, knowledge and skills, and standards			
Applied Learning • Use of technology • Selection of appropriate 21st-century skills • Ability of students to work well in groups and independently			
Active Exploration • Levels of student engagement • Ability of students to use inquiry skills and to think deeply • Student-centered instruction			
Adult Connections • Number of subjects, people, and organizations involved • Involvement of other adults • Adequacy of resources			
Assessment Practices • Enhancement of skill retention or standards mastery • Selection of culminating products and performances • Quality of rubrics • Quantity and mix of scaffolding and learning activities • Structured levels for students to self-assess their progress • Teacher's management of the process, coaching of students, and providing support			

es

A number of resources are available to help teachers design PBL mathematics units, some of which are listed in table 13.2. Some of these resources provide sample PBL units (from brief ideas to very detailed units). Others provide sample videos to support practitioners in implementing PBL units. Some resources showcase research studies that demonstrate the effectiveness of PBL, and others provide implementation tips and strategies for PBL practitioners. The resources checked in the final column, "Ideas That Drive Design," contain problems or challenges that may inspire a PBL unit topic.

Table 13.2. Some PBL resources

	Sample PBL Units	Sample Videos	Research	Tips for Using PBL	Ideas That Drive Design
Buck Institute for Education (http://www.bie.org)	✓	✓	✓	✓	
Project-Based Learning Handbook (Markham et al., 2003)	✓			✓	✓
PBL Starter Kit (Larmer et al., 2009)	✓			✓	✓
Edutopia (http://www.edutopia.org/project-based-learning)				✓	✓
PBLU.org http://pblu.org					
Indiana Collaborative for Project-Based Learning (http://www.rose-prism.org/moodle/prism/icpbl/)	✓				
MagnifyLearning (http://magnifylearning.org)	✓				
The PBL Academy (http://iuedmoodle.educ.indiana.edu/moodle/)	✓				
High Tech High (http://www.hightechhigh.org/projects/)	✓	✓	✓		
Curriki PBL Geometry (http://www.curriki.org/welcome/resources-curricula/curriki-geometry-course/)	✓				✓
Innocentive (http://www.innocentive.com)					✓
Mathalicious (http://www.mathalicious.com)					✓
Emergent Math (http://emergentmath.com/my-problem-based-curriculum-maps/)					✓
Real World Math (http://realworldmath.nctm.org)					✓

Embarking on this PBL journey is life changing. Shifting to a PBL approach can be a challenging endeavor for teachers, as it necessitates simultaneous changes in instructional materials and resources, pedagogical strategies, and assessment practices. For many teachers, it also disrupts their sense of identity, how they perceive themselves as teachers, and what their role should be in the classroom environment.

However, when done thoughtfully, PBL units in mathematics classrooms can become relevant and authentic experiences for both teachers and learners. Teachers scaffold content and activities in order to guide the learning process through a Driving Question. Learners actively construct knowledge in collaborative groups over an extended period of time to assist community partners. Posing provocative, open-ended Driving Questions and incorporating community partnerships are ways to make projects come to life and productively engage students in mathematics. How might you leverage PBL to continue your professional journey?

References

Colley, Kabba. "Project-Based Science Instruction: A Primer." *The Science Teacher* 75 (2006): 23–28.

Engle, Randi A. "Framing Interactions to Foster Generative Learning: A Situative Explanation of Transfer in a Community of Learners Classroom." *Journal of the Learning Sciences* 14 (2006): 451–498.

Henningsen, Marjorie, and Mary Kay Stein. "Mathematical Tasks and Student Cognition: Classroom-Based Factors That Support and Inhibit High-Level Mathematical Thinking and Reasoning." *Journal for Research in Mathematics Education* 28 (1997): 524–549.

Hiebert, James, Thomas P. Carpenter, Elizabeth Fennema, Karen C. Fuson, Diana Wearne, Hanlie Murray, et al. *Making Sense: Teaching and Learning Mathematics with Understanding.* Portsmouth, N.H.: Heinemann, 1997.

Larmer, John, David Ross, and John R. Mergendollar. *PBL Starter Kit: To-the-Point Advice, Tools and Tips for Your First Project in Middle or High School.* San Rafael, Calif.: Buck Institute for Education, 2009.

Schön, Donald A. *Educating the Reflective Practitioner: Toward a New Design for Teaching and Learning in the Professions.* San Francisco, Calif.: Jossey-Bass, 1987.

Smith, Margaret S., and Mary Kay Stein. *Five Practices for Orchestrating Productive Mathematics Discussions.* Reston, Va.: NCTM, 2011.

Smith, Margaret Schwan, Edward A. Silver, Mary Kay Stein, Melissa Boston, Marjorie Henningsen, and Amy F. Hillen. *Improving Instruction in Rational Numbers and Proportionality: Using Cases for Transforming Mathematics Teaching and Learning.* Vol. 1. New York, N.Y.: Teachers College Press, 2005.

Staples, Megan. "Supporting Whole-Class Collaborative Inquiry in a Secondary Mathematics Classroom." *Cognition and Instruction* 25 (2007): 161–217.

Stein, Mary Kay, Janine Remillard, and Margaret S. Smith. "How Curriculum Influences Student Learning." In *Second Handbook of Research on Mathematics Teaching and Learning*, edited by Frank Lester, pp. 319–370. Greenwich, Conn.: Information Age Publishing, 2007.

Stein, Mary Kay, Margaret Schwan Smith, Marjorie A. Henningsen, and Edward A. Silver. *Implementing Standards-Based Mathematics Instruction*. New York, N.Y.: Teachers College Press, 2000.

Stein, Mary Kay, Randi A. Engle, Margaret S. Smith, and Elizabeth K. Hughes. "Orchestrating Productive Mathematical Discussions: Five Practices for Helping Teachers Move beyond Show and Tell." *Mathematical Thinking and Learning* 10 (2008): 313–340.

Thornton, Holly. "Dispositions in Action: Do Dispositions Make a Difference in Practice?" *Teacher Education Quarterly* (Spring 2006): 53–68.